Identification of novel modulators towards high cell density and high-producing Chinese hamster ovary suspension cell cultures as well as their application in biopharmaceutical protein production

Dissertation

for the completion of the doctoral degree *doctor rerum naturalium*
at the technical faculty of the Bielefeld University, Germany

Submitted by

Beat Thalmann
(*16.01.1984)
Dipl. Chemist (FH), MSc Biotechnology (FH)
from
Berg am Irchel (ZH), Fischingen (TG) and Pfungen (ZH), Switzerland

2015

Chairperson: Apl. Prof. Dr. techn. Karl Friehs
1st Examiner: Prof. Dr. rer. nat. Thomas Noll
2nd Examiner: Prof. Dr. rer. nat. Kristian Müller
Associate examiner: Dr. rer nat. Heino Büntemeyer

Date of the oral examination: 16.06.2015

Bibliografische Information der Deutschen Nationalbibliothek

Die Deutsche Nationalbibliothek verzeichnet diese Publikation in der
Deutschen Nationalbibliografie; detaillierte bibliografische Daten sind
im Internet über http://dnb.d-nb.de abrufbar.

ISBN 978-3-8325-4046-3
ISSN 2364-4877

Logos Verlag Berlin GmbH
Comeniushof, Gubener Str. 47,
10243 Berlin
Tel.: +49 (0)30 42 85 10 90
Fax: +49 (0)30 42 85 10 92
INTERNET: http://www.logos-verlag.de

The laboratory work was fully established in the period between November 2008 and April 2013 at the Celonic GmbH (Jülich, Germany) according to the current state of the art. All cell lines that were used are properties of Celonic GmbH. The functionally revealed factors *cgr*Ttc36 and *cgr*Snord78 are patented:

Andreas Herrmann, Beat Thalmann, Arndt-René Kelter: Expression System. 05.11.2012; WO2014/068048 A1

Dedictated to my daughters Celia Félice and Lucia Florence....

"It's a magical world ... let's go exploring!"

Calvin and Hobbes, Bill Watterson, 31st December 1995

I Abstract

The present work covers the investigation of growth-promoting, cell density-increasing as well as productivity-enhancing cellular factors and the application in Chinese hamster ovary (CHO-K1) production cell lines. The freshly established serum-free progenitor CHO-K1-derived suspension cell line was chemically mutated achieving a randomly altered cell population, which enables the selection of cellular clones with superior growth characteristics. Selection at high osmolality (525 – 600 mOsmol·kg^{-1}), single cell cloning by limited dilution and clonal selection under harsh unregulated bioreactor conditions yielding in five high cell density-growing clones, of which one possessed a 2.5-fold increase in maximal cell density compared to control. The transcriptome analysis of the superior mutated CHO-S cell clone with the progenitor clonal control cell line and subsequent verification by quantitative real-time PCR resulted in six significantly deregulated transcripts. Hereafter, four transcripts (TPRp, LOC100759461, Gas5 and Ttc36) were validated regarding the influence on growth as well as production characteristics thereof by transfection of the progenitor and a hIgG-producing CHO-S cell line. Amongst these, *C. griseus* Ttc36 showed an immediate effect in increasing maximal as well as integral viable cell density. Surprisingly, the applied Gas5 variants did not show growth arrest, whilst the isolated over-transcription of intronic sequence possessing the C/D box snoRNA Snord78 actually exclusively increased the cell specific productivity and led to a 2.0 to 2.5-fold hIgG titer enhancement. In totally contrast to Snord78, Ttc36 overexpression did not affect cell specific productivity but the growth rate, cell density, cell aggregation as well as cell viability in a positive manner. Hence, two cellular factors with emerging positive impacts on biopharmaceutical recombinant protein production were revealed and successfully applied in early hIgG production processes. In this thesis, further insights in results and further discussions on putative mechanisms of these functional initially discovered cellular factors are provided.

II Zusammenfassung

Die vorliegende Arbeit umfasst die Detektion von wachstumsfördernden, die Zelldichte sowie Produktivität erhöhenden, zellulären Faktoren und deren Anwendung in CHO-K1 basierenden Produktionszelllinien (Ovarienepithelzellen von *Cricetulus griseus*). Eine daraus neu etablierte Serum-freie Suspensionszelllinie wurde durch chemische Mutagene genotypisch verändert, um eine zufällig veränderte Zellpopulation mit neuen, verbesserten Eigenschaften zu generieren. Zur Anreicherung von robusten Subpopulation sowie Einzelzellklonen wurde die mutierte Zellpopulation bei hoher Osmolalität (525 – 600 mOsmol·kg^{-1}), durch Einzelzellklonierung und klonale Selektion mittels repetitiven Langzeitkulturen selektiert. Daraus resultierten fünf CHO-S Zellklone mit erhöhtem und längerem Wachstum bei hoher Zelldichte, wobei eine klonale Zelllinie eine bis zu 2.5-fache maximale Zelldichte erreichte. Durch differentielle Transkriptomanalytik und nachfolgende Verifizierung mittels quantitativer Echtzeit-PCR von ebendiesem Zellklon sowie einer klonalen Kontrollzelllinie wurden sechs signifikant deregulierte Transkripte ermittelt. Basierend auf Transfektionsstudien und Überexpression in der Kontroll- sowie einer hIgG-produzierenden CHO-S Zelllinie wurden vier Transkripte (TPRp, LOC100759461, Gas5 and Ttc36) validiert. Unter allen Faktoren erwies die Überexpression von *C. griseus* Ttc36 sich als überaus vorteilhaft gleichermaßen in Bezug auf die Erhöhung der maximalen sowie integralen viablen Zelldichte. Anders als erwartet, verursachte die zelluläre Anhäufung der isolierten Gas5 Varianten keinen Wachstumsstopp, wobei die eigenständige Intron-basierte Transkription der Gas5-stämmigen C/D box snoRNA Snord78 eine drastische Erhöhung der Zell-spezifischen Produktivität und des hIgG Titers um das 2.0- bis 2.5-fache zur Folge hatte. Dem gegenüber führte die Überexpression von Ttc36 nicht zu einer Veränderung der Zell-spezifischen Produktivität, veränderte hingegen die Wachstumsrate, maximale sowie integrale Zelldichte, die Zellaggregation und die Viabilität im positiven Sinne. Somit wurden zwei zelluläre Faktoren mit positiven Eigenschaften für die Durchführung und Wirtschaftlichkeit der Produktion von biopharmazeutischen, rekombinanten Proteinen gefunden sowie erfolgreich auf ihre Anwendung in der Produktion von hIgGs getestet. Diese Dissertation gibt Einblicke in die Auswirkungen auf diverse CHO-S Zelllinien sowie mögliche Mechanismen dieser erstmals funktionell beschriebenen Faktoren.

III Contents

Contents

1 Aim of the work

In the following work, potential factors leading to high cell density growth or uncoupling of cell density dependent growth inhibition, respectively, should be discovered, verified and validated for their use in biopharmaceutical production cell lines.

There are different ways for manipulating cell lines to change their expression pattern and therefore for increasing the genomic variability for further selection procedures (Chapter 2.2). The method of choice should be the unspecific somatic mutagenesis of Chinese hamster ovary cells by various chemical compounds. Further, the mutated cells have to be cloned to produce single cell clones for subsequent selection of clones suitable to grow at high cell density.

Based on these high performance clones, transcriptomic as well as proteomic studies have to be performed, finding putative factors for uncoupling growth at high cell density. These factors (proteins, regulatory RNAs) should be verified by an independent method then. Finally, verified factors are validated by overexpression of suppression in precursor cell line using stable transfection of respective cDNA or stable expression of shRNA against these factors, respectively. By this step the number putative factors can be reduced and the effective effect on a cell line (here the precursor Chinese hamster ovary cell line) elucidated.

As a final step the residual factors need to be tested in producer cell lines to observe the behaviour in biopharmaceutical production and the capacity for increasing the high cell density in producer cell lines. The major step in this applied research project is crucial for the ability of a factor finding use in bulk biopharmaceutical production processes. Here, the product titer of the desired product must not be reduced by insertion or diminishment of high cell density factors.

Therefore, the main aim is to find at least one factor suitable to decouple the growth at high cell density and therefore increase the maximal cell density of a culture, but also the increase of product titer. The last point can be achieved if the specific productivity of each cell can be at least maintained.

2 Introduction

2.1 Short overview of Chinese hamster ovary cell lines

Chinese hamster ovary cell lines (CHO) were first published in December 1958 (PUCK *et al.* 1958) and were introduced as adherent cell line formerly isolated from a female donor of *Cricetulus griseus*. Following CHO cell lines are partially extensively used in biopharmaceutical production and research:

The CHO-K1 cell line was spontaneously immortalised and subcloned from the parental CHO cell line and possesses high capacity in growth as well as recombinant protein production. For recombinant protein production in CHO-K1 cell lines, antibiotic resistance gene for further selection of positively transfected cells are needed and therefore co-transfected with the gene of interest (GOI).

The Dhfr-deficient cell lines CHO-DXB11/CHO-DUKX (Urlaub and Chasin 1980) and CHO-DG44 (Urlaub *et al.* 1983) were both independently derived from chemically mutated parental CHO strain or the proline-deficient CHO-pro3⁻ strain, respectively. The CHO-DG44 cell line was further selected by methotrexate, subsequently γ-irradiated and again selected by tritiated deoxyuridine (Flintoff *et al.* 1976; Urlaub *et al.* 1983) and lacks in both allels of Dhfr gene. Both Dhfr negative cell lines are suitable for selection by insertion of exogenous Dhfr gene activity after transfection and multiple integration of the GOI coupled with the Dhfr gene. Selection is performed using medium in absence of glycine, hypoxanthine, and thymidine, whose are necessarily supplemented in growth media of these cell lines. Both cell lines were used for initial studies in stable transfection of mammalian cell lines towards biopharmaceutical protein production (Kaufman and Sharp 1982; Ringold *et al.* 1981). By applying high concentrations of methotrexate the copy number of an inserted Dhfr containing expression cassette can be increased. This method was lengthy suggested to be crucial for generation of high producer cell lines. However, recent studies showed no correlation between copy number and protein titer (Kwaks *et al.* 2003; Derouazi *et al.* 2006). Due to the elegant selection method for Dhfr positive cells and resulting high productivities, the CHO-DG44 cell line highly contributes to the total protein production in CHO derived cell lines (Jesus and Wurm 2011).

Both, CHO-K1 and CHO-DG44 are cell lines with high protein production abilities up to $5 - 10$ g·L^{-1} and possess suitable glycosylation patterns for human therapy (Butler and Meneses-Acosta 2012). In addition, CHO cell lines possess high growth rates and reach cell densities up to $1.2 \cdot 10^7$ cells·ml^{-1} in optimised batch cultures (Beckmann *et al.* 2012) or up to $2.5 \cdot 10^7$ cells·ml^{-1} (personal communications in industry and conferences).

Therefore, this shows the potential for high yielded biopharmaceutical glycoprotein production using CHO cell lines. The next chapters should highlight the genetic modification of Chinese hamster ovary cell lines using several possible techniques for improving their recombinant biopharmaceutical protein producing ability.

2.2 Unspecific mutations: A short survey of their use in cell line generation

Mutations can be caused spontaneously or directly mediated by irradiation, alkylating as well as intercalating chemical agents. Moreover, the integration of DNA fragments derived from retroviruses, transposons or artificial expression vectors within the genome were documented to induce alterations in genomic integrity as well (Stanford *et al.* 2001), which usually lead to point mutations, frame shifts or chromosomal translocations, in respective to the underlain mechanism (Thompson 1979).

Another method for inserting mutations is the passively mediated mutation using uncoupling the mismatch repair by deactivation of involved enzymes (Collins *et al.* 2011; Smith *et al.* 2005; Jenkins *et al.* 2011 pp. 115-119). This method increases the rate of spontaneous mutations ($5 \cdot 10^{-6}$ per locus, (Stanford *et al.* 2001), but needs the removal of the inserted effector elements (shRNA, siRNA or transposon).

The most convenient mutation methods are performed mutagenic substances initiating chemical modifications. Thereby, the mutagenic agents are simply added to the medium and removed after a defined incubation time by medium exchange. For physically driven mutations, there is need of a suitable irradiation source and equipment. Sole UV-irradiation by hand lamps are undefined and for γ-irradiation the proper equipment is necessary. Therefore, agents chemically inducing mutations are more cost effective and more efficient than physical sources (3- to 10-fold increased mutation rate, Stanford *et al.* 2001).

Engineered retroviruses, transposons and artificial expression vectors are insertional mutagens with higher specify and usually flanking homologous sequences. Artificial transposons are expression cassettes flanked by terminal inverted repeats and integration is mediated by the so called transposase, which is co-transfected (Ivics and Izsvák 2010). As long as the transposase is expressed, the transposonal expression cassette is excised and elsewhere integrated. This manner is advantageous over the simple expression cassette integration by DNA double strand break and followed non-homologous end-joining (Magin *et al.* 2013). At last, retroviruses (safety class 2 organisms) are difficult to handle as well as labour intensive and need additional safety requirements (Stanford *et al.* 2001; Ustek *et al.* 2012). All insertional mutagens have in common that they imply extensive cloning if compared with physical or chemical mutagens, whilst subsequent target finding is facilitated by sequencing starting from known sequences.

2.3 Selection methods for cell line generation of distinct phenotypes

As shown for the generation of CHO-DG44 cell line (Chapter 2.1), mutations lead to a wide variety of pheno- and genotypes, which have to be subsequently selected to gain the cell population with a desired function (Thompson 1979). This final selection is crucial for generation of cell lines with desired phenotypes and needs a specific selection pressure of physical or chemical manner. Again, for the CHO-DG44 cell line, the immortalised parental CHO cell line was mutated both by chemical as well as physical source and alternately selected by methotrexate (MTX) and tritiated deoxyuridine ([6-^{3}H]dUrd) respectively (Flintoff *et al.* 1976; Urlaub *et al.* 1983). Due to the residual Dhfr activity in MTX-resistant CHO cells, thymidine is still produced and incorporated into the DNA. In such cells, the selective agent [6-^{3}H]dUrd is converted to tritiated thymidine followed by DNA incorporation and cell death (Urlaub *et al.* 1981). Therefore, both selective agents were necessary to generate the diploidic Dhfr-negative CHO cell line DG44.

This example shows the necessity of a distinct and strong selection pressure. Other cell lines can be generated out of a mutated population by chemical, physical or environmental selection. All these processes involve the guarantee of a growth benefit for a certain cell subpopulation growing in a specific medium.

In contrast, complex selection pressures involve e.g. hypoxia (Zientek-Targosz *et al.* 2008) or hyperosmolality (Jenkins *et al.* 2011 pp. 115-119) and possibly benefit cells with not only one sole mutation. For instance, selection of e.g. suspension cell culture populations for hyperosmolality putatively isolates cells with increased resistance against autophagy/apoptosis (Han *et al.* 2010), improved salt homeostasis or both together. The modulation of apoptotic factors, autophagy and the related nutrient deprivation is a complex field itself and was deeply studied for enhancing cell line efficiency and biopharmaceutical product formation (Jeon *et al.* 2011; Dreesen and Fussenegger 2011; Dorai *et al.* 2010; Cost *et al.* 2010; Majors *et al.* 2009; Hwang and Lee 2009). In addition to cell death pathways, hyperosmolality can independently enhance expression of stress responding proteins, such as heat shock proteins (Lee *et al.* 2003) and osmolytes producing, toxic metabolite decomposing enzymes (Brocker *et al.* 2010; Brocker *et al.* 2011).

Another rather complex selection process is the adaption to hypothermic growth (Sunley *et al.* 2008). During continuously and repetitive growth at hypothermic condition a cell population is changed towards enhanced recombinant protein production, decreased volumetric cell density and maximal cell density as well as increased cell diameter due to higher mRNA content. Since a higher rate of nuclear fragmentation and multi-nucleation was observed, the present hypothermic adaption seemed to be both a genotoxic and selective process (Sunley *et al.* 2008).

In addition to the stress-related selection processes mentioned above, the desired cell line subpopulation could be isolated by repeated batch (Beckmann *et al.* 2012) and fedbatch

(Prentice *et al.* 2007) processes. This successive phenotype approximation by repeated identical cultivations usually leads to highly adapted cells. A simple repetition of batch cultures in a defined medium and subcultivation at mid-exponential growth phase lead to an adapted cell population with altered monoclonal antibody (mAb) production (Beckmann *et al.* 2012). During a continuing cultivation over 400 days, the clonal cell line derived from CHO DP-12 used in this study putatively changed its expression pattern by epigenetical modifications leading to total abolishment of hIgG heavy/light chain mRNA generation and mAb production (Osterlehner *et al.* 2011; Pilbrough *et al.* 2009). On the other hand, repetitive cultivations at exponential growth phase for up to 200 days showed no significant changes in stability and titer suggesting an overgrowth by an adapted subpopulation (Beckmann *et al.* 2012; Kaneko *et al.* 2010).

Nevertheless, repetitive batch cultivations cause a neither significant nor strong selection pressure for highly effective subpopulation generation. A long-term fedbatch end-point cultivation process showed better performance increasing the cell line parameters (Prentice *et al.* 2007). During each of four repetitive fedbatches, the cultures were cultivated until reaching viabilities below 50%. This process enabled the enrichment of a subpopulation with doubled maximal cell density, doubled chromosomal set as well as improved viability characteristics. Another more stringent repetitive fedbatch selection process also increased the product titer (Prentice *et al.* 2007).

Therefore, complex selection pressures are more difficult to handle with than constant selection pressures with simply one defined substance. On the other hand, complex selection pressures lead to highly adapted cell lines in normally inhibitory media.

More sophisticated methods are instrumentally driven sorting processes, which specifically enrich cell population with distinct features, such as antibody presentation (Böhm *et al.* 2004), fluorescence intensity/appearance (Zeyda *et al.* 1999; Kim *et al.* 2012) and metabolic properties (Hinterkörner *et al.* 2007; Bort *et al.* 2010). These methods are highly efficient for specific subpopulation enrichment from complex cell populations, but restricted in using the sorting systems FACS (fluorescence activated cell sorting) or MACS (magnetic activated cell sorting).

All selection processes need further analytics if the associated factors have to be identified. Such methods include traditional affinity and reporter assays as well as recently established genomic, transcriptomic and proteomic assays.

2.4 Identification of unknown regulatory cellular factors

Revealing unknown regulators and modulators of cell function, is prior made by affinity methods or differential expression profiling. Affinity methods need known biomolecules, DNA/RNA/protein fragments or small molecules and are primarily used for drug discovery against known cellular modulators instead of in unknown target identification thitherto. For early drug discovery known targeted biomolecule fragments and putative affinity binding pockets are needed (Bentley *et al.* 2013; Duong-Thi *et al.* 2013). For putative cellular functions or structure of unknown cell modulators, affinity methods could enrich such subpopulations for facilitating identification (Clark and Mao 2012).

But comparing mainly independent CHO cell lines once derived by somatic mutation and selection (Chapters 2.2 and 2.3), affinity based methods are not applicable, since usually no traceable residues or functions are known. Therefore, if control cell lines are available, differential expression profiling by proteomic or transcriptomic approaches are favoured. Due to the complexity of genomes and the complete unavailability of CHO genome sequences in recent past (Hammond *et al.* 2012a; Hammond *et al.* 2011; Xu *et al.* 2011), genomic comparison is not under consideration. Especially for comparing randomly mutated cell lines, genomic approaches may lead to poor results since putatively major changes are non-sense mutations with no impact in expression and activity. Dramatic changes such as translocations as well as single-allelic deletions surely have influence on overall expression ratios, but results are more diffuse and not transferable to neither mRNA or protein level (Geiger *et al.* 2010).

2.4.1 Proteomics

Using differential proteomic approaches, the effective modulators on protein level are compared. Here, the cell probes (supernatant or cell compartment lysates) of control and modified cell line are separately labelled, mixed and loaded on a 2D difference gel electrophoresis (2D-DIGE) gel. The proteins are subsequently separated regarding both isoelectric current as well as size (Meleady *et al.* 2011; Baik and Lee 2010). Hereinafter, differentially expressing protein spots are analysed by tandem mass spectrometric methods (e.g. ESI MS/MS, MALDI MS/TOF) and identified using peptide mass fingerprint databases (Shevchenko *et al.* 2001). In proteomic methods, the most limiting step are sensitivity and grade of separation in 2D-DIGE (Bunai and Yamane 2005). Membrane proteins as well as evanescent low amounts of transiently expressed regulatory proteins might be not efficiently displayed. In addition, a peptide mass fingerprint database for identification of dissected protein spots is needed (Aitken and Learmonth 2002; Cottrell 1994). For CHO, the experimental peptide mass fingerprints might differ from theoretically established databases of other species.

Another proteomic approach worth mentioning is called SILAC (stable isotope labeling with amino acids in cell cultures) (Grønborg *et al.* 2006; Liang *et al.* 2006; Colzani *et al.* 2008). Using this method, amino acids need to be labelled with stable isotopes (^{13}C, ^{15}N, D), exclusively added to a modified cell culture and subsequently applied to cultured cells using these heavy amino acids as building blocks. On the other hand, the control cell line is cultured by addition of standard media components (^{12}C, ^{14}N, H). After combination and tandem mass spectrometry the peptide mass fingerprints differ 1 amu each heavy isotop (^{13}C-^{12}C, ^{15}N-^{14}N, D-H).

Using on-line trypsination following one-dimensional gel electrophoresis (1-DGE), isoelectric focussing (IEF) and/or liquid chromatography coupled with electrospray ionization quadrupole ion-trap tandem mass spectrometry (LC-ESI-MS/MS), this discrimination enables relative quantification of expression without performing 2D-DIGE gels and manual spot dissection/trypsination (Giorgianni *et al.* 2003; Le Bihan *et al.* 2001; Liang *et al.* 2006).

2.4.2 Transcriptomics

Transcriptomic approaches such as cDNA microarrays and the next-generation sequencing methods are usually used for revealing of unknown RNA species in CHO cell lines including mRNA, ncRNA (non-coding RNA), miRNA (microRNA) as well as snoRNA (small nucleolar RNA) (Doolan *et al.* 2012; Doolan *et al.* 2010; Hammond *et al.* 2012b; Hackl *et al.* 2011).

Early comparative CHO transcriptomic analyses were performed using cross-species microarrays of *R. norvegicus* or *M. musculus* (Ernst *et al.* 2006; Trummer *et al.* 2008) or customised low-density CHO microarrays (Melville *et al.* 2011; Bahr *et al.* 2009) due to the officially unknown CHO-K1 genome and related sequence information until 2012 (Hammond *et al.* 2012a). In addition to this limitation and by applying mutated subpopulations, the hybridisation rate/efficiency of mutated cell line samples could differ to control cell line samples due to point mutations or truncated mRNA by induced alternative splicing as well as polyA tail synthesis.

Next-generation sequencing could be seen as a more convenient method, which can generate expression profiles based on RNA levels with higher coverage than cDNA microarrays (Birzele *et al.* 2010). Within this general term several sequencing techniques are combined, e.g pyrosequencing (Roche, 454 GS FLX), sequencing by synthesis (Illumina, HiSeq2000), sequencing by oligo ligation detection (AB, SOLiDv4) as well as dideoxy chain termination (Sanger 3730xl) (Liu *et al.* 2012). Here, sequencing by synthesis is the most cost effective method with high ability to perform high throughput analysis. Library preparation for the sequencing by synthesis method is performed starting at total RNA or mRNA following its fragmentation, cDNA conversion as well as 5'- and 3'-adaptor ligation. Alternatively, the isolated RNA is converted first and the resulting cDNA is fragmented afterwards. RNA fragmentation provide more even coverage along a transcript, whereat (oligo-dT) cDNA fragmentation lead to better reads at 3'-ends (Wang *et al.* 2009b). The resulting expressed sequence tags (EST) are further sequenced and aligned to a reference genome database. Through EST-genome alignment it is possible to define exonic and intronic regions (Trapnell *et al.* 2009). Furthermore, by prior isolation of small RNA species (e.g. miRNA) and next-generation sequencing (NGS) such species are selectively identified (Clarke *et al.* 2012; Hackl *et al.* 2011). Standard total RNA based NGS may lead in loss of small RNA species information due to the high EST number of each transcript.

Comparing all differential expression methods, the transcriptomic profiling by NGS leads to the highest coverage corresponding cross-species genome (Birzele *et al.* 2010) and allows analysis of small regulatory RNAs as well as mRNA (Clarke *et al.* 2012). The latter enables this method to be more powerful for determination of different expressed transcripts and for revealing of putative functions of hypothetical proteins than other comparable methods.

2.5   Pathway engineering towards high cell density cultures

The fade of mammalian cell lines in biopharmaceutical production cultivations is subdivided in four growth phases. Beginning at the consolidating lag phase, the cells start growing in an exponentially manner (exponential phase) until they reach the stationary and the final decline or death phase. Therefore, these four phases can be used for engineering and optimisation of a cell line (Dutton *et al.* 1998). Since the lag phase usually lasts about 24 hours, engineering at this phase seems not to be useful.

2.5.1   Cell cycle engineering

Increasing of growth rate within exponential phase leads to faster availableness of sufficient cell number for recombinant protein production, since at stationary phase the specific productivity is increased (Dean and Reddy 2013). Increased growth rate can be accomplished by cell cycle engineering using e.g. cMyc, mTOR, CDKL3 and E2F-1 overexpression. All reported cell cycle factors showed to increase the maximal cell density once overexpressed in CHO cell lines. A rather slight increase (20 %) of maximal cell density in batch cultures was observed overexpressing E2F-1 (Majors *et al.* 2008) and CDKL3 (Jaluria *et al.* 2007). Overexpression of cMyc (Kuystermans and Al-Rubeai 2009) and mTOR (Dreesen and Fussenegger 2011) in CHO cell lines lead to 2-fold increase in maximal cell density nevertheless starting at a low level. Furthermore, all cell cycle modulators above were commonly shown to shift the cell cycle towards higher S/G1 phase ratio.

Cell cycle engineering allows the generation of fast growing cell lines by uncoupling gate keeping mechanisms. Nevertheless, excessive growth stimuli could lead to increased necrotic cell ratio (Kuystermans and Al-Rubeai 2009). Here, modulation of cell death pathways could help to overcome this effect.

2.5.2   Modulation of cell death factors

The most convenient way to increase the integral viable as well as maximal cell density in CHO cell lines is the modulation of apoptotic pathways. Thereby, pro-apoptotic factor and anti-apoptotic factors are suppressed and overexpressed, respectively. Due to the complexity of the multiple programmed cell death pathways, production cell lines can be widely engineered (Krampe and Al-Rubeai 2010). All pathways (extrinsic, mitochondrial-intrinsic, ER-intrinsic and autophagy) were extensively studied and majorly lead to the final caspase-3/7 activation. The suppression of caspase-3 and caspase-7 lead to prolonged cultivation under cellular stress stimulation (Sung *et al.* 2007) and chemical inhibition of caspase-8 as well as caspase-9 enhanced cell viabilities in both batch and fedbatch cultures (Yun *et al.* 2007). Since the major apoptotic execution pathway was blocked this way, the cells possessed increased ability to be executed by autophagy (Sung *et al.* 2007).

On the other hand, overexpression of central pro-apoptotic proteins like Bcl-2 (Tey *et al.* 2000; Jeon *et al.* 2011) as well as Bcl-X$_L$ (Han *et al.* 2011; Kim *et al.* 2009c; Majors *et al.* 2012) in prolonged CHO cell culture for 1 – 3 days regarding viabilities over 80 % and significantly inhibited autophagy under extent nutrient deprivation as well as hyperosmolality (Han *et al.* 2011; Kim *et al.* 2009c). Therefore, these apoptotic regulators are suitable for application in fedbatch cultures, since autophagic and necrotic cell death is reduced, product degradation is avoided and downstream processing is facilitated (Kim *et al.* 2009c).

On the contrary, suppression of pro-apoptotic factors was described as well, whereby the most prominent execution proteins are the mitochondrially associated Bax and Bak proteins. Suppression of Bax and Bak by RNA interference caused prolonged viable cultivation about two days regarding viabilities over 80 % and resistance against cytotoxic stimuli (Lim *et al.* 2006). In addition, knockout of Bax and Bak clearly abolished apoptotic pathway onset by staurosporine and sodium butyrate addition as well as nutrient deprivation (Cost *et al.* 2010; Misaghi *et al.* 2013). This shows that total knockout of central pro-apoptotic factors is clearly more efficient than partial knockdown by small-interfering RNA (siRNA).

Whilst Bak/Bax induce intrinsic/extrinsic caspase cascade activation by mitochondrial cytochrom c release and Bcl-2/Bcl-X$_L$ inhibit this process, a group of BH3-only pro-apoptotic regulators transmit several stimuli to this proteins and stabilise or inhibit them, respectively (Kim *et al.* 2009a; Krampe and Al-Rubeai 2010). By suppression of BH3-only protein (e.g. Bid, Bim, and Puma) expression, Bak and Bax are not able to oligomerise under certain death stimuli (DNA damage, genotoxic stimuli, nutrient deprivation). Furthermore, overexpression of BH3-only pro-apoptotic proteins may lead to apoptosis even if Bcl-2 is overexpressed or Bax/Bak partially knocked down, as shown for Bim$_L$ (Plötz *et al.* 2013). In yeast, the extra-long isoform of Bim (Bim$_{EL}$) was shown to interact with Bax as well as Bcl-X$_L$, whilst Bim$_{EL}$ inhibited the anti-apoptotic ability of Bcl-X$_L$ but did not potentiate the pro-apoptotic Bax mechanism (Juhásová *et al.* 2011).

Other sensoric cell death proteins upstream of Bax/Bak and the Bcl-2 family proteins are Akt, as well as heat shock proteins (Hsp, chapter 2.5.3). The serine/threonine kinase Akt acts downstream of mTor (Chapter 2.5.1, Krampe *et al.* 2008) and several cell death stimuli such as nutrient/growth factor deprivation as well as DNA damage (Hemmings 1997). Therefore, overexpression of constitutively active Akt (CA-Akt) increased integral viable cell density in production CHO cell line batch cultures by resistance against nutrient deprivation in late stationary phase (Hwang and Lee 2009). In addition, CA-Akt overexpression inhibited autophagy as well, but could not increase the maximal cell density. Due to the prolonged culture duration, the antibody titer increased about 20 %.

On the other hand, downstream Bax/Bak activation other factors can be modulated avoiding further progress of apoptosis (Krampe and Al-Rubeai 2010). The overexpression of XIAP (Sauerwald *et al.* 2002; Kim *et al.* 2009b; Liew *et al.* 2010) and Aven (Chau *et al.* 2000) showed efficient inhibition of apoptosis induced by various stress stimuli in CHO and other production cell lines. Interestingly, the overexpression of XIAP and cultivation under standard condition without serum deprivation lead to rather lower cell densities than in cultures of the control cell line (Liew *et al.* 2010). In addition, Aven was shown to enhance the anti-apoptotic effects of the Bcl-2 homologues Bcl-X_L (Figueroa *et al.* 2004) and E1B-19K (Nivitchanyong *et al.* 2007; Dorai *et al.* 2009) once simultaneous overexpressed in CHO or BHK cell lines, respectively. In addition, CHO cell lines co-overexpressing E1B-19K/Aven and E1B-19K/Aven/XIAP, which are able to increase the lactate as well as ammonia/ammonium consumption, were shown to reach higher maximal cell densities in certain media and higher titers (approx. +60 %) in glucose-rich batch cultures (Dorai *et al.* 2009).

Taken together, modulation of cell death-associated factors has shown to prolong a protein production process efficiently by increasing the integral viable cell density. Nevertheless, no case mentioned above could repeatedly demonstrate enhancement of maximal cell density without chemical, physical or process-related stress application. Therefore, modulation of apoptosis as well as other cell death pathways is symptomatic and does not enable decoupling of the growth inhibition at high cell densities.

2.5.3 Relevance of heat shock proteins, molecular chaperons and detoxifying enzymes

In several proteomic as well as transcriptomic studies comparing exponential and stationary phase (Wei *et al.* 2011; Meleady *et al.* 2011) heat shock proteins (Hsp) and other chaperons were expressed in late culture phases. The appearance of PDI and other ER-associated chaperons at high cell density production cell line cultures is related to the recombinant protein overload within the endoplasmic reticulum (ER). The overexpression of the chaperone protein disulfide isomerase (PDI) in CHO production cell lines did not show clear growth beneficial effects, but increased the cell specific productivity (Mohan and Lee 2010; Davis *et al.* 2000). Comparing CHO cell lines with higher and lower growth rates, the molecular chaperon VCP (valosin-containing protein) (Wang *et al.* 2004) was found to impact cell density and growth (Doolan *et al.* 2010). Here, VCP was found to be overexpressed in fast growing CHO cell lines using both transcriptomic as well as proteomic methods. In three independent CHO production cell lines, suppression of VCP by siRNA led to reduced cell densities, whereat overexpression did not significantly increase the maximal cell densities of the cell lines. Hence, this example may show the non-trivial results in pathway engineering.

Overexpression of Hsp70 protein family members was found to beneficially alter growth and survival. The constitutive overexpression of human HSPA2 (HSP70-2) in V79 hamster fibroblast led to an increased resistance against chemically induced apoptosis (Filipczak *et al.* 2012). More recently, human HSP70 and HSP27 were overexpressed in CHO production cell lines (Lee *et al.* 2009). Here, HSP70 overexpression caused an approx. 185 % maximal cell density compared to the control cell line. In addition, the HSP70-overexpressing CHO cultures were able to maintain two days longer beyond viabilities of over 80 %. The overexpression of HSP27 did not lead to an increased maximal cell density, but enhanced the recombinant protein titer due to longer cell maintenance in culture. Respectively, HSP70 overexpression led to higher titers but decreased specific productivities. Interestingly, the constitutive co-expression of both HSP70 and HSP27 increased maximal cell density (1.4x), product titer (4x) and prolonged viable cell culture (+ 2 days, viabilities >80 %), combining both beneficial effects.

Therefore, HSP/Hsp proteins could be promising factors to increase the maximal viable cell density, the product titer and the resistance against apoptosis.

2.5.4 Hypothermic factors

Mild hypothermic cultures (30 – 34 °C) are widely used to increase the recombinant protein titer, the cell specific productivity by parallel reduction of cell density and to prolong the viable culture period (Kaufmann *et al.* 1999; Yoon *et al.* 2007). At hypothermic growth, the ratio of a cell population staying in G0/G1 cell cycle phase is increased (Fox *et al.* 2005). Therefore, growth arrest is initiated by mild hypothermic condition with increased ability for recombinant protein production per cell. Comparative studies were performed for elucidation of responsible factors leading to growth arrest and increased specific productivity, respectively (Yee *et al.* 2009; Kumar *et al.* 2008). Here, protein transport and modification transcripts were especially upregulated at hypothermic conditions, which supports the premise of increased cell specific productivities under hypothermia.

For instance, the cold-inducible RNA-binding protein (CIRBP/CIRBP) was found to be overexpressed in hypothermic BALB/3T3 mouse fibroblast cultures (Nishiyama *et al.* 1997) and during hypothermic adaption of CHO cell lines (Yoon *et al.* 2006). This cold-inducible protein is thought to impair growth even at standard culture temperature (37 °C), as elucidated in previous studies (Nishiyama *et al.* 1997). Interestingly, overexpression of CIRBP in adherent as well as suspension CHO production cell lines increased the production titer, but did not impair growth, cell density or viability (Tan *et al.* 2008). Furthermore, downregulation of CIRBP in CHO cell lines did not result in significant changes of cell density, product titer and growth at hypothermic as well as standard culture conditions (Hong *et al.* 2007). In contrast to BALB/3T3 mouse fibroblasts and other cell lines (Zeng *et al.* 2009), CIRBP had no impact on growth or survival in CHO cell lines.

Other CHO-specific factors with clear deregulation at hypothermic conditions are the microRNAs (miRNA) *cgr*-miR-7 and *cgr*-miR-21 (Barron *et al.* 2011; Gammell *et al.* 2007). The miRNA species *cgr*-miR-21 possessed no impact on growth or viability and slightly decreased specific productivity once overexpressed in CHO production cell lines (Jadhav *et al.* 2012). Whereas *cgr*-miR-21 is overexpressed under hypothermic conditions (Gammell *et al.* 2007), *cgr*-miR-7 is downregulated at stationary phase and significantly more downregulated after temperature shift from 37 °C to 31 °C (Barron *et al.* 2011). Here, knockdown of *cgr*-miR-7 by anti-miR-7 had no impact in cell density and protein titer at 37 °C as well as after temperature shift to 31 °C. Overexpression of pre-miR-7 decreased viable cell density of 37 °C culture to the level of temperature shifted culture, but the specific productivity increased significantly without increasing product titer. This observation overlaps with data from other cell lines (Xiong *et al.* 2011; Zhang *et al.* 2012), whereas overexpressed *cgr*-miR-21 was crucial for increase resistance against apoptosis and growth of various human cell lines ((Tao *et al.* 2011); (Gaur *et al.* 2011); (Wang *et al.* 2013c).

2.5.5 Metabolic engineering towards high cell densities

The classical pathway engineering is performed by modulation of metabolic pathways, to shift the equilibrium towards preferred metabolites. A common characteristic of CHO cell culture is the accumulation of the inhibitory metabolites lactic acid (lactate) as well as ammonia (ammonium) (Dean and Reddy 2013; Chen and Harcum 2005).

Lactate accumulation was shown to inhibit growth rate of CHO cell lines and affected cell viability at concentrations above 58 mM (Lao and Toth 1997; Xing *et al.* 2008). For reduction of lactate accumulation, the glycolytic pathway need to be shifted towards tricarboxylic acid cycle (TCA cycle) either by disruption of lactate formation or increase in lactate consumption. The lactate formation in CHO cell lines can be disrupted by knockdown of involved enzymes such as lactate dehydrogenase and pyruvate dehydrogenase kinase, which inhibits Acetyl-CoA producing pyruvate dehydrogenase (Zhou *et al.* 2011). Here, the lactate formation was effectively reduced, whereat the product titer was increased in parallel. Interestingly, because of enzyme knockdown the osmolality stayed stable, i.e. did not increase during cultivation. Furthermore, the integral viable cell density slightly but not significantly increased. A similar observation was made in other studies (Jeon *et al.* 2011; Kim and Lee 2007a). Here, lactate dehydrogenase A knockdown did not impair with growth or viability course, but reduced lactate formation significantly. Some clonal cell lines showed increased product formation possibly due to a more efficient energy homeostasis (Jeon *et al.* 2011).

The overexpression of pyruvate carboxylase in CHO cell lines is another way for reducing lactate accumulation by bypassing pyruvate decomposition through carboxylation to oxaloacetate and direct entry into TCA cycle (Kim and Lee 2007b). Again, overexpression of pyruvate carboxylase did not impair growth rate or maximal cell density, but slightly increased the life span of the cell culture. In contrast to lactate dehydrogenase knockdown, the overexpression of pyruvate carboxylase slightly reduced lactate formation only.

Overall, the modulation of lactate formation was slightly beneficial for culture longevity and product formation, but did not increase the maximal cell density or growth rate. On the other hand, overexpression of GLUT5 fructose-specific transporter in CHO cell lines and fructose feed during cell culture led to doubled viable cell densities as well as 6-fold reduced lactate formation in respective to glucose-feeded CHO/GLUT5 cultures (Wlaschin and Hu 2007). This example shows that the change of carbon source might increase maximal cell densities due to more efficient energy homeostasis. In succession, feeding of lactate instead of glucose led to decreased ammonium formation due to increased alanine production (Li *et al.* 2012). Nevertheless, the growth rate, viable cell density and culture longevity did not impair by lactate or pyruvate feeding.

Another way to decrease ammonium accumulation, which impairs cell growth above a concentration of 5.1 mM (Xing *et al.* 2008), was accomplished by overexpression of the taurine transporter (TAUT) (Tabuchi *et al.* 2010). In addition, TAUT overexpression led to reduced lactate formation and increased mAb (monoclonal antibody) titer due to a massively increased viable cell culture longevity. Interestingly, the overexpression of TAUT reduced the maximal cell density about 75 % in favour towards mAb production and specific productivity (Tabuchi and Sugiyama 2013). In addition, the enzyme alanine aminotransferase 1 (ALT1) was downregulated in TAUT overexpressed CHO cells. Thus, co-expression of TAUT with ALT1 increased mAb titer as well as viable cell density, whereas lactate and ammonium accumulation was reduced (Tabuchi and Sugiyama 2013). Concerning the enhanced cell density, the X-box binding protein-1 (XBP-1) was upregulated in ALT1 overexpressing CHO producer cell lines (Tabuchi and Sugiyama 2013). XBP-1 was shown to be activated at unfolded protein response (UPR), which usually occurs in highly productive mammalian cell lines (Shajahan *et al.* 2009). Hence, overexpression of the splice variant XBP-1s led to increased productivities in transient protein production (Ku *et al.* 2008) as well as stable adherent cell lines (Tigges and Fussenegger 2006). This coincidence of ALT1 overexpression, XBP-1 upregulation and increased viable cell density in TAUT overexpressing CHO cell cultures is possibly due to the unfastened UPR bottleneck and resulting growth impairment. In addition to lactate modulation by metabolic engineering, modulation of the ammonium metabolism did not show any beneficial effect on viable cell density.

If the limitation of high cell density growth is not based on formation of inhibiting metabolites, the cause could be underlain within the metabolic pathways. Nevertheless, by applying metabolomics on mAb-producing CHO cell lines malate was found to be extracellular accumulated during cultivation (Chong *et al.* 2010). This futile secretion pathway was deflected by overexpression of malate dehydrogenase II (MdhII) directing malate to oxaloacetate and TCA cycle. Overexpression of MdhII led to 1.4 – 1.9-fold improvement in integral viable cell density, 0.6-fold reduction in lactate as well as malate secretion rate and 1.2-fold increase in final mAb titer. In addition, this change of malate equilibrium increased the cell energy capacity, namely 3.5-fold/3.3-fold increased intracellular ATP and NADH concentration, respectively (Chong *et al.* 2010).

2.5.6 Increasing the maximal cell density by pathway engineering – A summary

Comparing all above discussed studies on pathway manipulation in CHO cell lines, the most efficient method to reach high cell densities up to $2.5 - 5.0 \cdot 10^7$ cell·ml^{-1} by fedbatch cultivation might putatively be accomplished by either stress resistance (Chapters 2.5.2 and 2.5.3) or functionally unknown proteins at the cell surface, which inhibit high cell density by cell-cell communication.

The modulation of cell cycle regulators and therefore the uncoupling of cell cycle in CHO cell lines did not lead to significantly increased maximal cell densities (Chapter 2.5.1). Furthermore, uncoupling of cell cycle rather led to increased cell death by necrosis (Kuystermans and Al-Rubeai 2009).

Modulation of cell death pathways as a logic consequence is highly efficient to generate robust CHO cell lines with prolonged viable cultivation duration (Chapter 2.5.2). Nevertheless, the majority of documented enhancements in maximal cell density were performed by application of cell death inducers or other cellular stresses.

During hypothermic cultures cell densities are significantly reduced in respective to cultures at standard conditions (Chapter 2.5.4). Here, the potential is not fully exploited; even the underlain mechanisms are different to high cell densities e.g. reached at perfusion cultures (Krampe *et al.* 2008). During such culture conditions the differential expression of Hsp family members are more likely, which showed significant increase in maximal viable cell densities, product titer and robustness once overexpressed in CHO cell lines (Chapter 2.5.3; Lee *et al.* 2009).

At last, custom metabolic engineering connected with initial metabolomics are emergent methods to efficiently generate and improve CHO high producer cell lines (Chapter 2.5.5). This approach surely possesses potential for increasing maximum cell density of a cell line and would lead to reduced nutrient limitation, but possibly not overcome the density-dependent growth inhibition.

In summary, the solution of uncoupling the density-dependent growth inhibition in CHO cells would be a combination of several methods including disruption cell-cell communication pathways, metabolic optimisation for proper nutrient supply at high cell densities and installing proper rescue pathways by apoptotic factors or Hsp family members. At the end, the most important point is to increase the overall product titer.

2.6   Towards high cell density growing CHO cell lines: A theoretical approach

For finding new factors, which inhibit or support growth at high cell densities, the generation of clonal cell lines with significantly increased maximal cell density is prerequisite. Subsequently, these clonal cell lines are needed for direct comparison with the precursor cell line.

First and foremost, the CHO-K1 precursor cell line needs to be genomically modified by random mutagenesis for variation of present phenotypes. Simple selection of present cell populations were not efficient enough to generate CHO cell lines with increased cell densities

(Bort *et al.* 2010; Taschwer *et al.* 2012). An elegant, fast as well as economical way for random mutagenesis is mediated by chemical agents (Chapter 2.2). A chemically mutagenic source would be more beneficial, since irradiation possesses a lower mutagenic efficiency compared with alkylating chemicals (Waldren *et al.* 1979; Keysar and Fox 2009). Due to the broad occurrence, the alkylating ethyl methanesulfonate (EMS) and the intercalating/alkylating acridine derivate ICR-191 have to be chosen for CHO cell line mutagenesis. EMS is an alkylating mutagen (Perez and Stamato 1999), which is used as standard mutagen for generation of mammalian cell lines (Chapter 2.2; Paulin *et al.* 1998) and plant cells (Rai *et al.* 2003) with stable phenotypes. Since EMS acts toxic as well (Perez and Stamato 1999; Sono and Sakaguchi 1988), the effective concentration should be elucidated first. In parallel, ICR-191 should be applied for CHO cell line mutation. This agent efficiently compromises alkylating and intercalating mutagenic capability (Fuscoe *et al.* 1979; O'Neill *et al.* 1978), acts directly on DNA and is less toxic for the cell. ICR-191 acts as frameshift agent and induces non-sense mutations and knockouts rather than point mutations observed with EMS (Ghaedi and Fujiki 2008; Zientek-Targosz *et al.* 2008).

After mutagenesis, the cells need to be initially selected for reducing clonal variation before performing single cell cloning. The applied selection pressure should simulate stresses at high cell density as well as increased osmolality (Chapter 2.3) and partial nutrient deprivation, e.g. by L-glutamine absence to shift to less futile metabolic pathways. In the following, a combination of all stresses should be employed without appearance of toxic metabolites, which usually occur in late phase fedbatches and would possibly negatively affect the selection. Therefore, the selection process should be performed with initial high cell density including continuing high osmolality (Jenkins *et al.* 2011 pp. 115-119) as well as absence of L-glutamine (Bort *et al.* 2010) and with culture end point at viabilities below 50 % (Prentice *et al.* 2007). These batch culture parameters should be maintained several passages without dilution for enrichment of robust subpopulations, which are easily able growing under these conditions. Both osmolality enhancement and cultivation to viabilities below 50 % led to cell line populations with increased maximal cell densities (Jenkins *et al.* 2011 pp. 115-119; Prentice *et al.* 2007). The absence of L-glutamine should select the cell population with reduced L-glutamine auxotrophy (Bort *et al.* 2010) and should simulate the absence of L-glutamine in late phase cultures.

After this initial selection, the putatively enriched cell population needs to be single cell cloned, whereat the resulting clonal cell lines are selected again. Within this selection process all clones should be cultivated in parallel and the presumably high number of individual clones should be reduced for further analytics. Since the readout for high cell density growth is insufficient, characteristics such as extensive cell aggregation as well as acidic medium colour should be employed for discrimination of dead or unprofitable cell clones. During selection the cell clones should be cultured using standard batch and fedbatch processes at initial high cell densities and long culture duration to enrich the most robust cell clones.

Once the clonal cell line with the best growth characteristics at high cell density is isolated, this mutated cell line is transcriptomically compared with a clonal cell line derived from the precursor CHO-K1 cell line with similar growth characteristics but the high cell density growth. The differential transcriptomic analysis preferentially performed by next-generation sequencing of the clonal cell lines leads to a set of differentially expressed factors (proteins, regulatory RNA) and needs independent analytics such as differential proteomic and quantitative PCR methods (Chapter 2.4) for verification of output results.

Once verified by independent methods, the factors are validated by stable transfection in the control precursor CHO-K1 cell line to elucidate the effect on high cell density growth during batch and fedbatch cultures. Furthermore, the pre-validated factors with cell density modulating abilities are transfected in clonal mAb-producing CHO cell lines to clarify the effect on volumetric and specific productivity as well as the interaction between high cell density and productivity. As a final step, the pool cell lines are separated by single cell cloning resulting in clonal cell lines, which are finally compared and characterised.

3 Material and methods

3.1 Cultivation

All cultivations were performed in CM1035 medium (HyClone Thermo, RR1302.01) supplemented with 4 mM L-glutamine (Gibco, 25030), unless described otherwise. All further supplements as well as exclusions were described separately (Table 3.1).

Table 3.1: Supplementation, feed media and alternate basal media

Product	Abbreviation	Source	Order number	Final concentration
F12 K Nutrient Mix (Kaighn's F12)	F12K	Gibco	21127-022	-
Fetal calf serum (NZ)	FCS	Gibco	16010	10 %
Geneticin	G418	Life Technologies	11811-031	400 $\mu g \cdot ml^{-1}$ †
Hygromycin B >85 %	Hygro	InvivoGen	ant-hm-5	235 $\mu g \cdot ml^{-1}$ (200 $\mu g \cdot ml^{-1}$)◊
Sodium chloride	NaCl	Roth	3957.1	100 – 137.5 mM †
Ascorbic acid	AscH	Roth	3525.1	5 $mg \cdot ml^{-1}$ †
CHO CD EfficientFeed A	EF A	Gibco	A10234	2.5 – 15 % ‡
CHO CD EfficientFeed B	EF B	Gibco	A10240	2.5 – 15 % ‡
IS CHO Feed-CD XP	IS	Irvine Scientific	91122	5 – 30 % ‡
CDM4PerMAb	-	Thermo HyClone	SH30871.02	-
CellBoost 6	CB6	Thermo HyClone	SH30866.01	2.5 – 17.0 $g \cdot l^{-1}$ †/‡
Recombinant human serum albumin, expressed in Rice	rHSA	Sigma	A9731-5G	1.5 $g \cdot l^{-1}$

† Solved in LAL Reagent Water (Lonza, W50-500) and pH adjustment if necessary prior use.
‡ Continuing addition in fedbatches leading to high theoretical concentration.
◊ Effective concentration

All cultivation containers used in this work are shown in table 3.2.

Table 3.2: Cell culture containers with appropriate volume range used for cultivation.

Container	Abbreviation	Source	Order number	Culture volume
Bioreactor filter tubes 50 ml	BFT50	TPP, Switzerland	87050	5 – 15 ml
Spinner flask 1000 ml	Sp1000	Schott, custom	n.a.	150 – 300 ml
Spinner flask 500 ml	Sp500	Schott, custom	n.a.	100 – 200 ml
Spinner flask 250 ml	Sp250	Schott, custom	n.a.	50 – 100 ml
T-75 flask	T75	TPP, Switzerland	90076	20 – 35 ml
T-25 flask	T25	TPP, Switzerland	90026	3 – 10 ml
6-well plates	6-well	TPP, Switzerland	92006	1 – 3 ml
12-well plates	12-well	TPP, Switzerland	92012	1 – 2 ml
24-well plates	24-well	TPP, Switzerland	92024	0.5 – 1 ml
96-well plates	96-well	TPP, Switzerland	92696	100 – 300 µl

All cultivations were performed at 37 °C $\pm$ 0.5 °C, 21 % O_2, 5 % CO_2 as well as >80 % humidity (except for Spinner flasks) and other container specific parameters (Table 3.3).

Table 3.3: Culture parameter for each culture container

Container	Agitation / Stirrer velocity	Airation	Amplitude
BFT50	200 rpm	A-D (A-E)	30 mm†
Sp1000	105 rpm	2.0 – 5.0 sL·h^{-1}	Approx. 30 mm‡
Sp500	105 rpm	1.0 – 2.5 sL·h^{-1}	Approx. 25 mm‡
Sp250	50-90 rpm	0.5 – 2.0 sL·h^{-1}	Approx. 20 mm‡

† Orbital shaker: Ovan Midi (Ovan, Spain) or SK-300 (JeioTech, Japan)
‡ depending on length and diameter of custom made magnetic pendulum as well as stirrer velocity

In addition, an empirical evaporation rate was determined, which differs from each culture container and culture condition (Table 3.4). For static (non-agitated) culture condition, no empirical evaporation rate was determined. All cell culture dependent parameters were calculated using the following equations in table 3.5.

Table 3.4: Mean empirical evaporation rate (MEER) for defined culture conditions. Standard conditions: 37 °C ± 0.5 °C, 21 % O_2, 5 % CO_2, >80 % humidity (except for Spinner flasks).

Container	Agitation / Stirrer velocity	Airation	Initial volume	Mean empirical evaporation rate †
BFT50	200 rpm	A-D	10 ml	0.1 ml·d^{-1}
Sp1000	105 rpm	2.0 sL·h^{-1}	200 ml	1.1 ml·d^{-1}
Sp500	105 rpm	1.0 sL·h^{-1}	100 ml	0.8 ml·d^{-1}
Sp250	90 rpm	0.5 sL·h^{-1}	60 ml	0.6 ml·d^{-1}

† It is likely that the change in agitation, aeration and volume can lead to other MEERs

3.2 Sampling

Cell suspension sampling out of cell culture containers was performed under sterile conditions. Depending on culture volume 50 – 500 µl cell suspension was removed (Table 3.6) and directly processed for cell counting (Chapter 3.3) as well as optionally for pellet/supernatant recovery (Chapter 3.2.1 and 3.2.2).

3.2.1 Cell culture supernatant recovery for analytical purpose

The appropriate cell suspension volume (100 – 900 µl) was centrifuged for two minutes at 10'000·g, room temperature (Biofuge pico, Heraeus). The supernatant was carefully removed from the pellet and was subsequently frozen at ≤-15°C.

3.2.2 Cell pellet recovery for RNA and genomic DNA isolation

The appropriate cell suspension volume carrying $2.5 \cdot 10^6 - 1.0 \cdot 10^7$ cells was centrifuged for five minutes at 200·g at room temperature (Biofuge pico or Megafuge, Heraeus). The resulting cell pellet was washed with 1 ml PBS (pH 7.4, 2-8°C, Gibco, 10010-049) and again centrifuged for five minutes at 200·g at room temperature (Biofuge pico, Heraeus). After removing the supernatant, the pellet was immediately frozen in liquid nitrogen and stored at ≤-65°C.

Table 3.5: Formulae for determination of cell culture dependent parameters

Cell density	
$$cd = n_{cells} \cdot f_{Dilution} \cdot 1250 \cdot ml^{-1}$$	n_{cells}: count cells using hemocytometer $f_{Dilution}$: dilution factor cd: cell density [cells·ml⁻¹]
Mean cell density	
$$\overline{cd} = \sum_{b=1}^{n} cd_b$$	b: number of biological replicates cd_b: individual cell density $\overline{cd}$: mean cell density
Standard deviation †	
$$S_{cd} = \sqrt{\frac{\sum_{j=1}^{n}\left(cd_j - \overline{cd}\right)}{n-1}}$$	n: sample size (biological replicates) S_{cd}: Standard deviation of the biological replicats
Viability	
$$viability = \frac{vcd}{tcd}$$	vcd: viable cell density tcd: total cell density
Integral viable cell density (icvd)	
$$ivcd = \sum_{j=1}^{n}\left(vcd_{j-1} \cdot \Delta t_{j-1} + \frac{\left(vcd_j - vcd_{j-1}\right) \cdot \Delta t_j}{2}\right)$$ $$j=1: \; ivcd = \frac{vcd_1 \cdot \Delta t_1}{2}$$	Δt: time difference between sampling j: day j
Doubling time (DT)	
$$DT = \frac{\ln(2) \cdot \left(t_{j+1} - t_j\right)}{\ln\left(\frac{vcd_{j+1}}{vcd_j}\right)} = \frac{\ln(2)}{\mu}$$	t_j: sampling time at day j vcd_j: viable cell density at day j μ: growth rate
Feed / Supplement addition	
$$V_F = V_A \frac{c_B}{c_F - c_B}$$ $$V_B = V_A + V_F = V_A\left(1 + \frac{c_B}{c_F - c_B}\right)$$	V_F: volume of stock solution added V_A: Culture volume before feed addition V_B: Culture volume after feed addition c_A: desired feed concentration after feed addition c_F: feed concentration in stock solution

† Equation suitable for all parameters

Table 3.6: Cell culture containers with appropriate volume range used for cultivation.

Culture vessel	Sample volume range	Standard sample volume
BFT50	50 – 500 µl	100 µl (batch) / 300 µl (fedbatch)
Sp1000	300 – 1000 µl	500 µl
Sp500	100 – 1000 µl	500 µl
Sp250	100 – 500 µl	300 µl
T75	100 – 500 µl	300 µl
T25	100 – 1000 µl	100 µl
6-well	100 – 200 µl	100 µl
12-well	50 – 100 µl	100 µl
24-well	50 – 100 µl	50 µl
96-well	n.a. †	n.a. †

† No cell sampling applicable/performed

3.3 Manual cell counting

All cell densities were estimated by sampling the appropriate volume of cell suspensions (50 µl – 100µl) and manual counting with hemocytometer (Neubauer improved, Brand). By default, the samples were incubated with the same volume (50 µl – 100µl) of trypsin 0.5 %/EDTA (Gibco, 10x) for 3 – 5 minutes at 37 °C (water bath). Subsequently, the trypsinated samples were diluted with PBS 1x (Biochrom AG) / erythrosine B (Roth) (3:1 – 8:1) to achieve the appropriate cell density for cell counting, i.e. to reach a cell density of $1.125 \cdot 10^5 - 2.5 \cdot 10^5$ cells·ml^{-1} (or in terms of 8x 1 mm^2 hemocytometer chambers: 90 – 200 total counts). Through addition of erythrosine B, viable cells can be distinguished from cells that lost their cell membrane integrity and were supposed to be dead. Here, erythrosine B passes through the disintegrated cell membrane of dead cells, which are stained therefore. Viable cells do not show staining.

3.4 Cryopreservation and thawing

The appropriate cell suspension volume carrying $5.0 \cdot 10^6 - 1.0 \cdot 10^7$ cells each cryo was centrifuged for five minutes at 209·g at room temperature (Megafuge, Heraeus). The pellet was cryopreserved in x ml freezing medium (for x cryos) containing 45 % fresh and 45 % conditioned medium (from current cultures) and 10 % DMSO (Sigma, D2650). Homogeneously resuspended cells were aliquoted à 1 ml per vial (Cryo tubes, TPP, 89020) and transferred into Nalgene cryopreservation boxes (cooling rate 1 °C·min^{-1}). These boxes were immediately stored at ≤-65 °C for at least 10 hours up to seven days before the vials were transferred into the vapour phase of a liquid nitrogen tank (≤-150 °C).

For starting a culture, cryopreserved cell lines were directly thawed in room tempered (20 – 30 °C) cell culture medium (1 ml cell cryo cell suspension + 9 ml medium). The resulting suspension was centrifuged for five minutes at 209·g and room temperature, following resuspension in desired cell culture medium volume at room temperature for immediate cultivation.

3.5 Next generation sequencing

For identification of differentially transcribed RNA species, cell pellets in biological independent triplicates were shipped on dry ice to DNAVision S.A. in Belgium for RNA isolation, gene amplification, clustering and final sequencing (NGS: next generation sequencing) using Illumina HiSeq2000 and the appropriate kit systems. Reads were aligned against the mouse reference genome (NCBIM37.60) and the identified transcripts were compared to those annotated with the reference genome.

3.6 Quantitative real-time PCR analysis

3.6.1 Total RNA isolation

The total RNA was isolated from freshly prepared or frozen cell pellets (Chapter 3.2.2) by NucleoSpin RNA II Kit (Macherey-Nagel, 740955.50, auxilliar 2-mercaptoethanol: Sigma, 63689) according to the manufacturer's manual. The absorption of the final total RNA solution was measured at λ = 260 nm (Abs_{260nm}) (Ultrospec 1000E, Pharmacia Biotech) and subsequently diluted with LAL Reagent Water (Lonza, W50-500) to 200 ng·μl^{-1}.

3.6.2 cDNA generation

For complementary DNA (cDNA) generation, 2 µg (10 µl) of total RNA (Chapter 3.6.1) and 1 µl primer (25 pmol or 200 ng, respectively) (Table 3.7) were incubated for five minutes at 70 °C and for five minutes at 4 °C. Subsequently, 10 nmol (1 µl) dNTPs (NEB, N0447S), 200 U (1 µl) M-MuLV RT Polymerase (NEB, M0253S), 1x (2 µl) M-MuLV RT Buffer (10x, NEB, B0253S) as well as 5 µl LAL Reagent Water (Lonza, W50-500) were added to obtain 20 µl total volume.

Table 3.7: Primers for cDNA synthesis

Primer	Sequence	Source	bp	Use
poly dT primer (25 µM)	TCGAGCGGCCGCCATGTGTTTTTTTTTTTTT TTTTTT	TIP custom	36	Standard primer
random Hexamer (200 ng·µl⁻¹)	NNNNNN	Fermentas SO142	6	applied if spliced introns were needed to be co-amplified

The mixture was incubated for one hour at 37 °C and heat inactivated for 10 minutes at 70 °C. The resulting cDNA (theoretically 100 ng·µl^{-1}) was used directly or stored at ≤-15°C.

3.6.3 Quantitative real-time PCR analysis

Each quantitative real-time PCR reaction (qRT-PCR) was performed within Strip Tubes 0.1 ml (LTF-Labortechnik GmbH, Wasserburg) in technical duplicates and 20 µl total reaction volume using the Rotor-Gene 2000 system (Corbett Research).

For 20 µl reaction mixture, 10 ng template (cDNA in respective to total RNA concentration) and 1 pmol each primer (Table 3.8) (both in 5 µl LAL Reagent water each) were added to 10 µl SensiMix^{TM} SYBR No-Rox Kit 2x (Bioline, Luckenwalde, QT650-02) at room temperature. In addition to the potential high cell density factors, the internal standard actin B (ActB) was always measured for each template as well as run. Run-to-run fluctuation/variation was reduced therefore.

Table 3.8: Primers for quantitative real-time PCR analysis (Forward: fw; Reverse: rv). Primers were designed on strong homologous sequences of *M. musculus*, *R. norvegicus* and *H.hapiens* homologues or on basis of sequenced hamster (*cgr*) data. The primers are aligned in chronological order. TPRp, GTPase and PPase are anonymised, whereat primer sequences are not listed.

Primer	Sequence	bp	T$_M$ [°C]	Target
ActB_fw	TCATGTTTGAGACCTTCAACACCCCAGCC	29	70.1	Actin B (Standard)
ActB_rv	GGTCCCGGCCAGCCAGGTCCAGACGCAGG	29	82.0	
AI661_fw	CAACTCCTCGAGACACCCGGTGACTGG	26	69.5	AI661453
AI661_rv	GGGCCGGCTACCTCTGTAAAAGATGC	26	68.0	
Ank1_fw	CCATGTATGGAGGCACCATTATACCCTCTC	30	65.7	Ankhd1
Ank1_rv	TTTGGCGCGCCGTTGACATATTTGAGATGC	30	74.0	
BC068_fw	CCACTCCATAAGCTTTGCAGCTTAAATGC	29	65.3	BC068281
BC068_rv	TTGAAATGTATGTGAGTCAGATCCAGAGG	29	63.9	
Btg_fw	CCTCTTCCAATGTGGCACCCTTTGCCC	27	72.8	Btg3
Btg_rv	TCCAGGAGGAGGTACCCATGTCACTGG	27	68.3	
C22Rik_fw	CACTGCCCAGCAGGAACTGAGCCTGC	26	71.1	5330417C22Rik
C22Rik_rv	CCAACAGCTGTATAAATGTGGTCACCAGC	29	66.7	
Cdkc_fw	CAGCGCCTGCCGCAGCCTCTTCGGGCC	27	82.7	Cdkn1c
Cdkc_rv	CAGCGCCCCACCTGCACCGTCTCGCGG	27	83.0	
Ebf1_fw	ATGAAGGAAGAGCCGCTGGGCAGCG	25	69.5	Ebf1
Ebf1_rv	CCCTGTCTGTCGTAGAGGGCCAGGACG	27	72.6	

Primer	Sequence	bp	T_M [°C]	Target
Filip1_fw	CAACCTTTGCCAGAGCACAGACCCCAGAG	29	70.9	Filip1l
Filip1_rv	TGAAACTGCCATGATGACTGCCGGTCTGG	29	69.5	
Gas5_fw	CTTGTAACCAGTGATGTGATGATTCTGCC	29	65.3	Gas5
Gas5_rv	GCATGTCCACTTGTCACAGGAGCCCTAAG	29	69.5	
cgrGas5_fw	GCAGAAATGGACTTGCAATTCAGTC	25	61.3	Specific for cgrGas5
cgrGas5_rv	AGAGAGTTCAAGTTGTGGTGAATCATCAC	29	63.9	
GTPase_fw	n.a.	28	71.0	GTPase†
GTPase_rv	n.a.	29	70.9	
hIgG_LC_fw	ACTGCCTCTGTTGTGTGCCT	20	59.4	hIgG light chain
hIgG_LC_rv	GCTGTAGGTGCTGTCCTTGC	20	61.4	
hIgG_HC_fw	CAAGTGCAAGGTCTCCAACA	20	57.3	hIgG heavy chain
hIgG_HC_rv	ACCAGACAGGTCAGGGACAC	20	61.4	
Hmga1_fw	ACTGAGAAGCGAGGCCGGGGCAGG	24	71.3	Hmga1
Hmga1_rv	CAGTTTCTTGGGTCTGCCCCTTGGTTTCC	29	69.5	
Hygro^R_fw	ATGAAAAAGCCTGAACTCACCGCG	24	65.8	Hygromycin B resistance gene
Hygro^R_rv	CCACGCCCTCCTACATCGAAGCTG	24	67.9	
I04Rik_fw	ATGGTGCGCTGGCCAGGCCTGAGG	24	71.3	A930005I04Rik
I04Rik_rv	AAGAACTCCTCCCAGACCAGCATGAGGCC	29	70.9	
ΔIntron_fw	GCGTCTACGGTGGGAGGTCTATTAAAGC	28	90.0	Fully spliced IntronA or Intron_cgrSnord78
ΔIntron_rv	TGGAGCCAAACGCAGTACAAAGTGTTACC	29	90.0	
J02Rik_fw	GCATCCAGATGTACTTCAAAGTGAAGGCG	29	66.7	9330101J02Rik (LOC100759461)
J02Rik_rv	AACACCAAGAAGTAATACCTCGGTTCTCC	29	65.3	
K11Rik_fw	CGCCGCGTCTTTGCGGTGGAGC	22	69.6	9930012K11Rik
K11Rik_rv	GGAAGCCAGTTTCAGAAGTGGCATCTGTC	29	68.1	
L06Rik_fw	GAATTCTGCAGTGGATTGCTCTGGCCTGC	29	69.5	9330182L06Rik
L06Rik_rv	TTCATCAAATTTGATGCCACTGCCCAAGG	29	65.3	
mi137_fw	GGCCCTCTGACTCTCTTCGGTGACGGG	27	72.6	mir-137
mi137_rv	TGCCGCTGGTACTCTCCTCGACTACGCG	28	73.4	
Pcdc_fw	AAGTTCTGGATCAGTCAGCCCGGGCCAG	28	73.3	Pcdc5
Pcdc_rv	CCATTACTTTTCTTCTGTTGAATTTCACTG	30	60.3	

Primer	Sequence	bp	T_M [°C]	Target
PPase_fw	n.a.	30	70.8	PPase†
PPase_rv	n.a.	30	70.3	
TPRp_fw	n.a.	30	71.3	TPRp
TPRp_rv	n.a.	25	72.8	
*cgr*Snord78_fw	GGAATCCAACACCGGCTATAGGCCAG	26	68.0	*cgr*Snord78
*cgr*Snord78_rv	CTTAGTTGCCTGTCACAGGATTCTTCAG	29	65.3	
Ttc36_fw	ATGGGGACTCCAAATGATCAGGCAGTG	27	66.5	Ttc36
Ttc36_rv	TCACTCCCTGCAACTCCAGGGCCTTGG	27	71.0	
Vezt_fw	GTGTGAAAGTGGGGCTGAAT	20	57.3	Vezatin
Vezt_rv	GTTCCTGCATGGTGGTGAAT	20	57.3	
Yaf_fw	CATCCATGGCAAAGCTGTAA	20	55.3	Yaf2
Yaf_rv	TTTGGTAAAAGGCCAATGAA	20	51.2	

† TPRp, PPase and GTPase are anonymous replacements for known proteins

Table 3.9: Parameters/settings during real-time PCR reaction

Step	Temperature	Duration	Purpose	Cycle(s)
Activation	95 °C	12 min	Initial heat activation of polymerase	1
Amplification	95 °C	25 s	Template denaturation	45 – 55 ‡
	55, 60 or 65 °C †	15 s	Primer annealing	
	72 °C	30 s	Extension/Elongation	
Melt	60 °C ---> 90°C	10 min	Generation of a melting curve (quality control)	1
Hold	40°C	2 min	Cool down the system	1

† Annaeling temperature depended on primer melting temperature (lowest $T_m > T_{annealing}$)
‡ Cycle repetition depended on fluorescence culmination of all samples

In following the Ct value (cycle threshold) for each sample was determined by metering the cycle number at logarithmic normalised amplitude fluorescence of 0.100 ± 0.001. All Ct values were dedicated as raw data. Before using the raw data for further calculation, the melting curves of a target were interpreted to be coherent. Not coherent samples were rejected.

The difference in expression was determined and calculated using the ΔΔCt method (Schefe *et al.* 2006) (Table 3.10). In addition, the final standard deviation was calculated using the law of error propagation (Nordgård *et al.* 2006) in comprehension of mean values as well as standard deviation of both controls and samples.

Table 3.10: Formulae for quantitative real-time PCR parameter determination

Mean	
$$\overline{Ct} = \sum_{b,t=1}^{n} Ct_{bt}$$	b: number of biological replicates t: number of technical replicates Ct: cycle number at logarithmic normalised amplitude fluorescence of 0.10
Standard deviation	
$$S = \sqrt{\frac{\sum_{j=1}^{n}\left(Ct_j - \overline{Ct}\right)}{n-1}}$$	n: sample size (biological and technical replicates)
Differential expression parameters †	
$\Delta\Delta Ct = \left(Ct_{gene,control} - Ct_{ActB,control}\right) - \left(Ct_{gene,sample} - Ct_{ActB,sample}\right)$ $\Delta\Delta Ct = \Delta Ct_{control} - \Delta Ct_{sample} = \log_2\left(fold\ change\right)$ $fold\ change = 2^{\Delta\Delta Ct}$ † $\Delta\Delta Ct > 0$: Overexpression; $\Delta\Delta Ct < 0$: Suppression	gene: the variable target ActB: Actin B sample: mutated or transfected sample control: identically cultured control sample (untreated, Mock)
Error propagation	
$$E_{\Delta\Delta Ct} = \frac{1}{\overline{Ct}_{control}^2}\sqrt{\overline{Ct}_{sample}^2 \cdot S_{\overline{Ct}_{control}}^2 + \overline{Ct}_{control}^2 \cdot S_{\overline{Ct}_{sample}}^2}$$ $$E_{foldchange} = \ln(2) \cdot E_{\Delta\Delta Ct} \cdot 2^{\Delta\Delta Ct}$$	$S_{\overline{a}}$: Standard deviation of technical and biological replicate mean

3.7 Preparative Methods

3.7.1 Genomic DNA isolation

The genomic DNA was isolated from freshly prepared or frozen cell pellets (Chapter 3.2.2) by NucleoSpin Tissue Kit (Macherey-Nagel, 740952.10) according to the manufacturer's manual. The absorption of the final genomic DNA solution was measured at $\lambda = 260$ nm (Abs_{260nm}) (Ultrospec 1000E, Pharmacia Biotech) and optionally diluted with LAL Reagent Water (Lonza, W50-500).

3.7.2 Preparative PCR

Preparative PCR was performed for amplification of target genes out of cDNA, gDNA or pDNA (complementary, genomic or plasmid DNA). cDNA was synthesised using protocol in chapter 3.6.2. Genomic DNA was isolated using procedure at chapter 3.7.1. All reactions were performed in PCR tubes (SoftTubes 0.2 ml, Biozym, 711080) with following content (Table 3.11).

Table 3.11: PCR mixture for preparative amplification.

Component	Source	Product number	Final Concentration	Volume added
Phusion Buffer HF 5x	NEB	B0518S	1x	20 µl
dNTP Mix, 10 mM each	NEB	N0447S	200 µM each	2 µl
Forward Primer, 100 µM	MWG Eurofins	custom	1 µM	1 µl
Reverse Primer, 100 µM	MWG Eurofins	custom	1 µM	1 µl
Template (cDNA, gDNA or pDNA)	C. griseus	-	20 pg·ml^{-1} to 1 µg·ml^{-1}	1 – 10 µl
Phusion Hot Start Flex DNA Polymerase, 2 U·ml^{-1}	NEB	M0535S	20 U·ml^{-1}	1 µl
LAL Reagent Water	Lonza	W50-500	-	ad. 100 µl

The real-time PCR reaction is performed using following settings in table 3.12 in consideration to the melting temperatures of the primers.

Table 3.12: Parameter/setting overview for preparative PCR

Step	Temperature	Duration	Purpose	Cycle(s)
1	98 °C	65 s	Initial heat activation of polymerase	1
2 †	98 °C	25 s	Template denaturation	0 – 10 ‡
	65 – 67 °C	15 s	Primer annealing	
	72 °C	30 – 100 s ◊	Extension/Elongation	
3	98 °C	25 s	Template denaturation	30 – 50 ‡
	72 °C	30 – 100 s ◊	Extension/Elongation	
4	72 °C	10 min	Completion of PCR	1
Hold	4 °C	∞	Cool down the system	-

† Initial annealing since primer coverage is approx. 70% according original template.
‡ Depending on primer coverage and template amount.
◊ Depending on fragment size

For each target, the specific parameters were applied as shown in table 3.13. The corresponding primer pairs for target amplification are shown in table 3.14. The target LOC100759461 could not be amplified. Therefore, LOC100759461 was synthesised then (Chapter 3.7.3).

Table 3.13: Target specific parameters/settings of successful preparative PCR

Target	Size ◊	Template †	Runs ‡	Annealing		Elongation Duration	Total Cycles
				Temperature	Cycles		
TPRp	400 bp	cDNA (10 ng)	1	65 °C	5	45 s	35
Ttc36	600 bp	cDNA (10 ng)	1st	65 °C	5	45 s	35
		1st PCR (2pg)	2nd	-	-	60 s	40
Gas5	3000 bp	gDNA (100 ng)	1	67 °C	5	100 s	45
Gas5 spliced	1000 – 3000 bp	cDNA (100 ng)	1	67 °C	5	100 s	45
Snord78	481 bp	pDNA (CV143, 100 pg)	1	60 °C / 66°C	10 / 35	30 s	45

◊ Supposed size through homology with *M. musculus*/*R. norvegicus*
† All templates derived from CHO source; cDNA: complementary DNA; gDNA: genomic DNA; pDNA: plasmid DNA
‡ For enrichment further runs were performed

Table 3.14: Preparative PCR primer for amplification of target genes

Primer †	Sequence (5'--->3')	bp	T_M [°C]	Description
TPRp_fw	AAACGCGTGCCACCATGxxxxxxxxxxxxxxxxx	34	71.9	Amplification of *C. griseus* TPRp (*cgr*TPRp)
TPRp_rv	GCCTACTAGxxxxxxxxxxxxxxxxxxxxxxxx	29	70.9	
Ttc36_fw	AAACGCGTGCCACCATGGGGACTCCAAATGATCAG	35	73.0	Amplification of *C. griseus* Ttc36 (*cgr*Ttc36)
Ttc36_rv	CACTAGTCAGCGCCCGTTACTGGGCGCGC	29	> 75	
*cgr*Gas5_fw	ATCGCCTGGAGACGCCACCAGGTAACAGGGGCAGG	35	>75	Amplification of *C. griseus* Gas5 variants (*cgr*Gas5) out of genomic and complementary DNA
*cgr*Gas5_rv	TAAGTGAGAGAGTTCAAGTTGTGGTGAATCATCAC	35	67.1	
*cgr*Snord78_fw	AAACCTAGGTTGAGTAAGTATTGGAATCCAACAC	34	65.9	Amplification of *C. griseus* Snord78 Intron (*cgr*Snord78) out of *cgr*Gas5
*cgr*Snord78_rv	AAACTCTAGAGTACTCCTACCTAAATAAAATAGG	34	63.4	

† fw: forward primer; rv: reverse primer

3.7.3 Custom gene synthesis

The hamster gene LOC100759461 (mouse homologue 9330101J02Rik) was externally synthesised by GeneArt (Life Technologies) with prior codon optimisation for *C. griseus* and/or manual restriction site masking. The delivered GeneArt vectors (5 µg) were solved in 50 µl LAL Reagent Water (Lonza, W50-500) for at least two hours at room temperature or for 16 – 18 hours at 2 – 8 °C and were used for restriction digest directly (Chapter 3.7.4).

3.7.4 Restriction digest

In general, procedures for plasmid DNA restriction were performed according to procedures published earlier (Sambrook *et al.* 2001). For individual enzymes, restriction digests were performed according to manufacturer's protocol (NEB) in terms of incubation conditions (buffer, time, temperature) and heat inactivation. Restriction digests were performed for analytical as well as preparative purposes. All restriction enzymes and corresponding buffers were purchased from NEB. In general, $1 - 10$ U·µg^{-1} DNA each enzyme and restriction site were used for restriction digest (Table 3.15). The reaction was incubated for $1 - 2$ hours at 37 °C (other enzymes used $55 - 65$°C). If applicable, the enzymes were heat deactivated for 20 minutes at $65 - 80$°C or directly cleaned up by agarose gel electrophoresis (Chapter 3.9.2) and/or NucleoSpin® Gel and PCR Clean-up Kit (Macherey-Nagel, Düren, 740609.50) for further processing (Chapter 3.7.7). For analytical digests (Table 3.16), 20 µl reaction volume with 0.1 – 0.5 U enzyme was applied. Due to the reduced enzyme concentration, the reaction was prolonged to $2 - 3$ hours.

Table 3.15: General restriction digest set up for cloning experiments

Component	Final Concentration	Volume added †
pDNA/PCR fragment	$50 - 500$ ng·µl^{-1}	x µl
NEBuffer 1-4, 10x	1 x	5 µl
Restriction enzyme (each restriction site)	$1 - 10$ U·µg^{-1} DNA	$0.5 - 4$ µl each enzyme
BSA, 100X (optional)	1 x	0.5 µl
LAL Reagent Water (Lonza, W50-500)	-	ad. 50 µl

† Standard volumes. Buffer, BSA (optional) and LAL Reagent Water can be adjusted to higher reaction volumes

Table 3.16: General restriction digest set up for analytical purpose

Component	Final Concentration	Volume added
pDNA (MiniPrep)	approx. 20 ng·µl^{-1}	1 µl
NEBuffer 1-4, 10x	1 x	2 µl †
Restriction enzyme	$5 - 25$ U·ml^{-1}	$1 - 5$ µl·ml^{-1} †
BSA, 100X (optional)	1 x	0.2 µl †
LAL Reagent Water (Lonza, W50-500)	-	ad. 20 µl †

† applied as freshly premixed mastermix

Subsequently, the restriction pattern was observed using agarose gel electrophoresis and further graphical documentation (Chapter 3.9.2).

3.7.5 DNA dephosphorylation

To avoid self-ligation and to reach higher probabilities for positive clones after ligation, vector backbones were dephosphorylated with antarctic phosphatise (NEB, M0289S) following instructions of the manufacturer (NEB). Antarctic phosphatase buffer (NEB, 0289S) was diluted in the restriction sample to 1x concentration. The mixture was then incubated for one hour at 37°C and heat inactivated for 10 minutes at 65°C or directly purified (Chapter 3.7.7).

3.7.6 DNA phosphorylation

Blunt PCR fragments generated by Phusion Hot Start Flex DNA Polymerase (Chapter 3.7.2) lack on 5'-phopsho residues for proper ligation process (Chapter 3.7.8). Therefore, purified PCR fragments were treated with $5U/\mu g_{DNA}$ T4 polynukleotide kinase (PNK, NEB, M0201S) in 1x T4 DNA Ligase Buffer (NEB, B0202S). The reaction mixture was incubated for one hour at 37 °C. Furthermore, the enzyme could be inactivated for 20 minutes at 65°C or was directly purified (Chapter 3.7.7).

3.7.7 Isolation and Purification of DNA Fragments

Purified DNA fragments were obtained by purification from resulting pieces of agarose gel electrophoresis or directly from PCR reactions using NucleoSpin® Gel and PCR Clean-up Kit (Macherey-Nagel, Düren, 740609.50) according to manufactures manual. Alternatively, elution was performed with LAL Reagent Water (Lonza, W50-500).

3.7.8 Ligation

Ligations of DNA fragments (dephosphorylated backbone vector and phosphorylated inserts) were performed using 400 U (1 µl) T4 DNA ligase (NEB, M0202S) in 20 µl total volume with 1x T4 DNA Ligase Buffer (2 µl) (NEB, B0202S). Molar ratio of backbone vector to insert was 1:3 or 1:3:3 if two inserts were applied together, respectively. Typically, 100 – 200 ng total DNA was ligated, whereat to constant backbone DNA amount the necessary amount of insert DNA to reach desired molar ratio was added. Incubation was performed for 16 hours at 16 °C with subsequent heat inactivation for 20 minutes at 65°C.

3.7.9 Transformation of competent *E.coli* DH5α cells

Frozen (≤-65°C) and chemically competent *E.coli* DH5α cells (NEB 5-alpha Competent E. coli (High Efficiency), NEB, 100µl) were thawed on ice for 10 minutes and combined with DNA solutions. From each ligation-reaction 5 µl and for purified supercoiled plasmid DNA 25 ng DNA were used, respectively. After incubation on ice for 30 minutes, bacteria were subjected to a heat-shock at 42°C for 30 seconds. After cooling on ice for 5 minutes, 900 µl SOC media (animal component free, NEB, Kit component) without antibiotics were added and incubated for one hour at 37 °C and 600 rpm (amplitude: 3 mm). After incubation, each transformation (50 – 200 µl bacterial suspension) was plated on LB-agar-plates (containing 100 µg·ml^{-1} ampicillin) (Chapter 3.7.10, Table 3.17). Incubation was performed at 37 °C overnight. Plates with defined colonies were used further directly or sealed and stored at 2 – 8 °C.

3.7.10 *E.coli* cultivation

Transformed *E.coli* DH5α cells were cultivated either on LB-Amp plates or in LB-Amp medium both containing 100 µg·ml^{-1} ampicillin (Table 3.17). LB-Amp plates were solely used for clone generation and selection after transformation (Chapter 3.7.9).

Cultures for analytical purposes and clone screening were performed in 3 ml LB media + 100 µg·ml^{-1} ampicillin (LB-Amp medium) for 16-18 hours at 37 °C and 220 rpm (MiniPrep, chapter 3.7.11). Inoculation was performed by adding a single plate colony.

Larger cultures were inoculated with 50 – 100 µl MiniPrep culture and performed in 100 ml LB-Amp medium within a sterile 500 ml Erlenmeyer flask (Schott) for 16-20 hours at 37 °C and 180 rpm (MidiPrep, chapter 3.7.11). The culture duration depended on the optical density at λ = 600 nm (OD$_{600nm}$), which must not be exceeded OD$_{600nm}$ = 4.0 (OD$_{600nm}$ $\leq$ 4.0) and was measured in semi-micro cuvettes (Roth, Y199.1). The plasmid bearing cells were harvested by centrifugation for 15 minutes at 4 °C and 4260·g.

Table 3.17: Formulations for LB medium and plates containing ampicillin

	Component	Source	Product number	Concentration	Procedure
LB Medium	Sodium chloride	Roth	9265.1	10 g·l^{-1}	All components were solved in Milli-Q water (18.2 MΩ·cm) and pH was adjusted to 7.0. Sterilisation for 20 minutes at 121 °C.
	Yeast extract	Applichem	A3732.0100	5 g·l^{-1}	
	Soy peptone	Applichem	A2206.0250	10 g·l^{-1}	
LB plates	Agar-Agar, bacteriogical	Roth	2266.1	15 g·l^{-1}	In combination with LB medium. Added before Sterilisation and subsequently plated with prior ampicillin addition.
	Ampicillin (·Na)	Roth	K029.1	100 mg·l^{-1}	Added to sterile and low tempered LB medium or LB agar plate solution.

3.7.11 Plasmid DNA preparation

Preparation of plasmid DNA from small culture volumes for analytical purposes (MiniPreps) was performed using the NucleoSpin Plasmid Kit (Macherey-Nagel, Düren, 740588.250) according to manufacturer's instructions. The washed plasmid DNA (pDNA) was eluted with 60 µl elution buffer (AE, 5 mM Tris/HCl pH 8.5). Further, the resulting pDNA-containing solutions were digested with suitable restriction enzymes (Chapter 3.7.4) for obtaining a specific pattern (fingerprint) once applied on an agarose gel to identify the correctly assembled plasmids (Chapter 3.9.2).

Plasmid DNA preparation for transfection and cell line generation purposes (Chapter 3.8) was performed using the Nucleobond Xtra Midi EF kit (Macherey-Nagel, Düren, 740420.10) according to manufacturer's instructions. After elution, 3.5 ml 2-propanol (AppliChem, A0900.2500PE) was added. For proper precipitation, the solution/suspension was incubated for five minutes at ≤-15 °C and subsequently centrifuged for at least 60 minutes at 4 °C and 4260·g. After precipitation, all procedures were performed under sterile conditions. The DNA pellet was washed twice with sterile 70 % ethanol (Roth, 9065.2) in LAL Reagent Water (Lonza, W50-500) and dried for at least 30 minutes at room temperature (20 – 30 °C). The pellet containing the desired plasmid DNA was solved in 100 µl LAL Reagent Water.

Quantification and quality check was performed by measurement of the UV absorbance of diluted plasmid DNA samples at 260 and 280 nm. For measurement of the optical density of plasmid DNA, disposable UV micro cuvettes (Roth, Y201.1) were used.

Quantity and quality of plasmid DNA are calculated according Table 3.18. If the quality did not meet ($R_{Abs} > 2.10$; $R_{Abs} < 1.80$), the plasmid DNA was precipitated (Chapter 3.7.12) or produced freshly again.

Table 3.18: Formulae for quantitative real-time PCR parameter determination

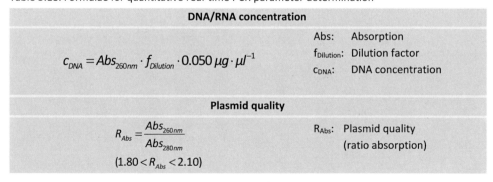

DNA/RNA concentration	
$$c_{DNA} = Abs_{260nm} \cdot f_{Dilution} \cdot 0.050\,\mu g \cdot \mu l^{-1}$$	Abs: Absorption $f_{Dilution}$: Dilution factor c_{DNA}: DNA concentration
Plasmid quality	
$$R_{Abs} = \frac{Abs_{260nm}}{Abs_{280nm}}$$ $$(1.80 < R_{Abs} < 2.10)$$	R_{Abs}: Plasmid quality (ratio absorption)

3.7.12 Linearisation of vectors and sterile DNA precipitation

For transfection in 25-CHO-S derived cell lines, all delivering plasmids were prior digested with 1 U·µg$_{DNA}$$^{-1}$ BspHI (NEB, R0517S) in 1x NEBuffer 4 (NEB, B7004S) for one hour at 37 °C to obtain linearised vectors with excised ampicillin resistance gene. Further, the fragments were unsewn by agarose gel electrophoresis (Chapter 3.9.2), from which the linearised vector was excised and purified by NucleoSpin® Gel and PCR Clean-up Kit (Chapter 3.7.7). Elution was always performed with 100 µl NE Buffer (Manufacturer's standard protocol 30 – 50 µl).

The resulting unsterile vector fragment solution (100 µl) was precipitated by adding 10 µl Sodium acetat (Roth, 6773.1) (3 M, pH 5.2) and 220 µl ethanol (Roth, 9065.2) (≤-15 °C) for 16 – 8 hours at ≤-15 °C or 5 – 10 minutes at ≤-65 °C. The precipitated DNA was pelleted by centrifugation for 60 – 90 minutes at 10'000·g and 4 °C.

After precipitation, all procedures were performed under sterile conditions. The DNA pellet was washed twice with sterile 70 % ethanol (Roth, 9065.2) in LAL Reagent Water (Lonza, W50-500) and dried for at least 30 minutes at room temperature (20-30 °C). The pellet containing the desired plasmid DNA was solved in 10 – 30 µl LAL Reagent Water (depending on the amount of digested vector).

DNA concentrations were measured (Chapter 3.7.11) and volume adjusted with LAL reagent water get reach 100 pg·µl^{-1}·bp^{-1}· (e.g. 0.5 µg·µl^{-1} for a 5000 bp fragment).

3.8 Vectors and cell line generation

The following, materials and methods regarding cell line development are described.

3.8.1 Overview on basal vectors and their derivatives

Two basal vectors (CV001 and CV121) were used for subcloning as well as vector construction respectively (Tables 3.19 and 3.20).

CV121 derived from CV001 by replacement of pA$_{BGH}$-ori$_{f1}$-P$_{SV40}$-Neo^R through IRES$_{FMDV}$-EM7-Hygro^R. This new vector was suitable for ensuring the coupled expression of the gene of interest (GOI) and the hygromycin B resistance gene (Hygro^R). CV001 was exclusively used for subcloning, since *Sac*II as well as *Bsm*BI restriction sites were unique compared to CV121 for removing the IntronA sequence. The basal vector CV121 was either modified within the multiple cloning site (MCS) for generation of an expression cassette (P$_{CMV}$ -IntronA-<u>GOI</u>-IRES$_{FMDV}$-Hygro^R-pA$_{SV40}$), by replacing the present IntronA by one or more intronic sequences (IOI: Intron of interest) (P$_{CMV}$-IOI-MCS-IRES$_{FMDV}$-Hygro^R-pA$_{SV40}$) or by vector fusion (addition of expression cassette) (P$_{CMV}$-GOI-pA-P$_{CMV}$-IntronA-MCS-IRES$_{FMDV}$-Hygro^R-pA$_{SV40}$). In table 3.21, all CV121 derivates, which were designed and constructed in this project are shown. The basal vector CV001 was exclusively used for subcloning, since other unique restriction sites were available (*Sac*II and *Bsm*BI respectively).

Table 3.19: Basal vector CV001 for subcloning and final vector construction.

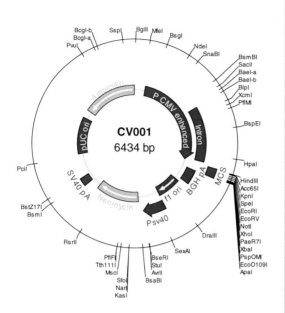

P$_{CMV\ enhanced}$:	P$_{CMV}$ with enhancer and IntronA (P$_{CMV}$-IntronA)
Intron:	IntronA of bovine serum albumin (IntronA)
MCS:	Multiple cloning site
BGHpA:	Bovine growth hormone poly-adenylation signal (pA$_{BHG}$)
f1 ori:	f1 phage origin of replication (ori$_{f1}$)
P$_{SV40}$:	Simian virus 40 promoter
Neomycin:	aminoglycoside phosphotransferase gene (Neo^R)
SV40 pA:	Simian virus 40 poly-adenylation signal (pA$_{SV40}$)
pUC ori:	pUC plasmid origin of replication (ori$_{pUC}$)
Ampicillin:	β-lactamase (Amp^R)

Table 3.20: Basal vector CV121 for final vector construction and its special features.

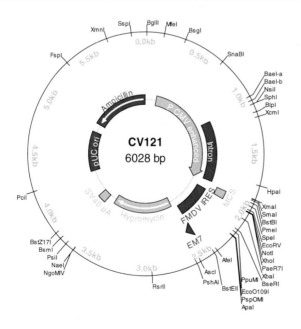

FMDV IRES:	Internal ribosome entry site of foot-and-mouth disease virus (IRES$_{FMDV}$)
EM7:	synthetic T7 promoter
Hygromycin:	Hygromycin B phosphotransferase gene (Hygro^R)

3.8.2 Nucleofection of CHO-S cell lines

CHO-S cells for nucleofection (transfection by Amaxa/Lonza Nucleofector II) were cultivated at least three passages before nucleofection. Exponentially growing cell cultures were employed for nucleofection. Hence, cell culture cell densities between 1.0 and $3.0 \cdot 10^6$ cells·ml^{-1} are typical for the exponential growth phase. The required cell culture volume was transferred to a 50 ml centrifuge tube (NerbePlus, 02-572-7001) and centrifuged for five minutes at 209·g and 20 – 30 °C. The supernatant was discarded and the pellet was resuspended in a defined amount of room tempered Nucleofector solution V (20 – 30 °C, from Amaxa Cell line Nucleofector Kit V, VCA-1003), so that a final cell density of $2.0 \cdot 10^6$ cells·ml^{-1} was reached. For every approach, 100 µl cell suspension in Nucleofector solution V ($2.0 \cdot 10^6$ cells) were mixed with 1000 pg·bp^{-1} linearized plasmid DNA (max. 10 µl) (Chapter 3.7.12) within a Nucleofector cuvette (from Amaxa Cell line Nucleofector Kit V, VCA-1003), which then was positioned in the Nucleofector II and pulsed with U-024 (formerly optimized for this cell line). After the nucleofection, the cells were immediately resuspended in 2 ml of room tempered (20 – 30°C) medium and transferred into a 6-well plate (TPP, 92006), resulting in a cell density of $1.0 \cdot 10^6$ cells·ml^{-1}. The cells were incubated at 37 °C, 5 % CO_2, 21 % O_2 and > 80 % relative humidity.

Table 3.21: CV121 derivatives. All unique vector elements (UVE) were flanked by CV121 vector elements CMV and IRES$_{FMDV}$-EM7-Hygro^R. In case of CV142-CV144, the UVE was inserted between Amp^R and CMV of CV121. The vector CV139 (CV121-based) possesses a synthetic intron flanking an AvrII restriction site downstream SacII (double cutter in CV121). UVE and respective backbones are flanked by applied restriction sites.

Vector	Size / bp	Unique vector elements (UVE)	Vector backbone
CV142	8955	$^{BglII}P_{CMV} - cgr$Gas5 $-pA_{SV40\ BstZ17I}$	$^{MfeI(Klenow)}$CV121$_{BglII}$
CV143	8163	$^{BglII}P_{CMV} - cgr$Gas5$_{Snord\Delta47\Delta77/80\Delta79} - pA_{SV40\ BstZ17I}$	$^{MfeI(Klenow)}$CV121$_{BglII}$
CV144	7714	$^{BglII}P_{CMV} - cgr$Gas5$_{Snord\Delta47\Delta77/80\Delta78\Delta79} - pA_{SV40\ BstZ17I}$	$^{MfeI(Klenow)}$CV121$_{BglII}$
CV145	6385	^{PmeI}cgrTPRp$_{SpeI}$	^{SpeI}CV121$_{PmeI}$
CV146	6572	^{PmeI}cgrTtc36$_{SpeI}$	^{SpeI}CV121$_{PmeI}$
CV147	7558	^{PmeI}cgrTPRp $-$ IRES$_{FMDV}$-EM7$_{AscI} - {}^{MluI}cgr$Ttc36$_{SpeI}$	^{SpeI}CV121$_{PmeI}$
CV155	5536	$^{AvrII}cgr$Snord78$_{XbaI}$	^{SpeI}CV139$_{AvrII}$
CV156	6140	$^{AvrII}cgr$Snord78 $_{PmeI} - {}^{PmeI}cgr$Ttc36$_{SpeI}$	^{SpeI}CV139$_{AvrII}$
CV159	9306	^{XmaI}LOC100759461$_{XbaI}$	^{XbaI}CV121$_{XmaI}$

3.8.3 Selection of positively transfected cells and further cultivation

The selection of positively transfected cells started 18 – 24 hours after transfection by adding 200 µg·ml^{-1} Hygromycin B (>85 %, InvivoGen, ant-hm-5) as selection pressure. During the following cultivation, every third to fourth day 1 ml selection medium was added and every seventh to eighth day the medium was exchanged by centrifugation of cells for five minutes at 209·g and 20 – 30 °C. Here, the supernatant was discarded and the cells resuspended in 2 ml fresh culture medium containing 200 µg·ml⁻¹ Hygromycin B. In course of the selection, positively transfected cells could survive under these conditions only. Typically around 11 – 14 days post transfection, the overall viability of the nucleofection approach reached a minimum, whereat recovery was observed after 17 – 21 days. The resulting clone pool was further kept under selection pressure. After the clone pool reached viabilities of 90 – 100 % the cells were cryopreserved (Chapter 3.4).

3.8.4 Limited dilution and clonal cell line generation

Early or late exponentially growing cells from clone pool cultures ($1.5 - 5.0 \cdot 10^6$ cells·ml⁻¹) were diluted to a final concentration of $3 - 30$ cells·ml⁻¹ in cultivation medium (37 °C) (Table 3.22) and seeded in a 96-well plate (TPP, 92696) with 100 µl cell suspension per well (0.3 – 3.0 cell per well). After seeding, each well was observed within up to 22 hours for single cells. These wells were documented and marked properly to ensure retrieval of wells with single cell. To avoid extended cooling, only half plates were observed at once, which took about 40 – 50 minutes.

Table 3.22: Limited Dilution media and set up

Cell line	Medium	Supplementation	Seeding cell number
CHO-S-E400$_{HyOsm}$	CM1035	4 mM glutamine 10 % IS CHO Feed-CD XP	1 cell·well⁻¹ (12 plates)
25-CHO-S	CM1035	4 mM glutamine	1 cell·well⁻¹ (2 plates)
	CM1035	4 mM glutamine 25 % PBS 1x	1 cell·well⁻¹ (2 plates)
	CM1035	4 mM glutamine	0.5 cell·well⁻¹ (2 plates per cell line, total 6 plates) †
23-CHO-S/CV107/K20-3/ CV146lin ‡	CM1035	5 mM glutamine 10 % IS CHO Feed-CD XP	1 cell·well⁻¹ (6 plates)
	CM1035	5 mM glutamine	0.3 cell·well⁻¹ (21 plates)
	ProCHO5	4 mM glutamine	1 cell·well⁻¹ (1 plate)
	ProCHO5	4 mM glutamine 1.5 g·l⁻¹ rHSA	3 cell·well⁻¹ (1 plate)
23-CHO-S/CV107/K20-3/ CV155lin ‡	ProCHO5	4 mM glutamine 1.5 g·l⁻¹ rHSA	3 cell·well⁻¹ (6 plates)

† second limited dilution with single cell cloning
‡ without 600 µg·ml⁻¹ G418 / 200 µg·ml⁻¹ Hygromycin B

After seven days, 100 µl cultivation medium was added to reach approx. 200 µl final volume. Again seven days later, a further addition of 100 µl cultivation medium was performed. Usually, the total cell suspension volume was 220 – 250 µl (depending on position) after three weeks.

After two weeks, the 96-well plates were microscopically analyzed and monitored for the confluence of the growing clonal subpopulations. Conglomerations of single cell clones with a confluence ≥ 50% were transferred to 6-well plates (TPP, 92006) for further cultivation (with 600 µg·ml⁻¹ G418 / 200 µg·ml⁻¹ Hygromycin B for transfected cells) and cryopreserved.

The efficiency of a Limited Dilution is calculated as followed (Table 3.23).

Table 3.23: Equation for Limited Dilution efficiency

Limited Dilution efficiency (E_{LD})	
$$E_{LD} = \frac{n_{survived}}{n_{single\,cells}}$$	$n_{survived}$: number of clones in continuous culture $n_{single\,cells}$: number of wells with single cells

3.9 Analytics

3.9.1 Fixation of cells and DAPI staining

The respective cells suspension was centrifuged (Chapter 3.2) and washed with 1 ml PBS pH 7.4 (Gibco, 10010-049) (2 – 8°C). The resulting cell suspension in 1 ml PBS was put in a 22.1 cm² tissue culture dish (TPP, 93060) and 110 µl of 37 % formaldehyde solution (Roth, 4979.1) resulting 4 % formaldehyde suitable for fixation was added. The fixation reaction was performed for 15 minutes at 2 – 8 °C. Afterwards, the cells were washed twice with 2 ml PBS pH 7.4 (2 – 8°C) and floated with 1 ml PBS 1x pH 7.4 (2 – 8°C). Then, 1 µl of 5.0 mg·ml⁻¹ DAPI (4′,6-diamidino-2-phenylindole) stock solution (14.3 mM, Life Technologies, D1306) was added and incubated for 30 minutes at 2 – 8 °C. Finally, the cells were washed twice with 2 ml PBS pH 7.4 (2 – 8°C), floated with 1 ml PBS 1x pH 7.4 (2 – 8°C) and observed under 20x magnification using the appropriate UV source and filter for DAPI (excitation/emission maxima: 358/461 nm). After fixation, ≥400 cells were counted and indexed in fragmented (>1 nuclei per cell) and single nuclei.

3.9.2 Agarose Gel Electrophoresis

For confirmation of identity (by size, restriction fragment pattern), quality and quantity, agarose gel electrophoresis of digested plasmid and PCR fragments was performed with agarose gels containing 0.8 – 1.0 % agarose (AppliChem, A2114.0500) in 1x TBE buffer (50 ml or 100 ml; Table 3.24) in the presence of SYBR Safe 1x (Invitrogen, S33102) for 90 minutes at 110V. For small fragments (100 – 400 bp) a high percentage agarose gel (2 – 4 %) was applied.

Gels were prepared in a way to avoid an overload of DNA per lane, so max. 1 – 2µg per lane were tolerated, otherwise slots were combined. This case was important for preparative agarose gels (Chapters 3.7.2 & 3.7.7).

Table 3.24: Components for 5x TBE agarose gel electrophoresis buffer[‡].

Component	Source	Final concentration	Mass †
Tris	Roth 5429.1	450 mM	109.0 g
Boric acid	Roth 6943.1	450 mM	55.7 g
EDTA·2Na·2H$_2$O	Roth 8043.3	10 mM	7.45 g
Milli-Q Water	-	-	ad. 2.0 l

† Components were mixed together and adjusted to pH 8.3 ± 0.1 before completion to 2.0 L
‡ 1x TBE buffer was made by dilution of 5x TBE buffer in Milli-Q Water

For proper application, the appropriate volume of 5x DNA loading buffer (Table 3.25) was added to the DNA samples to reach 1x buffer concentration (restriction digest: 20 – 100 µL; 5x DNA loading buffer: 5 – 25 µl). Subsequently, the resulting samples were applied to the agarose gel.

Table 3.25: Components for 5x DNA loading buffer

Component †	Source	Final concentration	Mass ‡
Tris·HCl (1M, pH 8.0)	Roth 5429.1	25 mM	250 µl
EDTA (0.5M, pH 8.0)	Roth 8043.3	150 mM	3.0 ml
Bromophenol blue (·Na) (0.1 %)	Roth A512.1	0.025 %	2.5 ml
Glycerol	Roth 3783.1	25 %	3.15 g
Milli-Q Water	-	-	ad. 10.0 ml

† All solutions were prepared before using Milli-Q Water and sterile filtration (0.22 µm)
‡ Completed buffer was sterile filtered (0.22 µm). Storage at 2 – 8 °C

For DNA fragment size and concentration estimation, 10 µl 2-log DNA ladder was added to each agarose gel. For known band pattern and/or preparative agarose gels, no 2-log DNA ladder was added if desired. For gel documentation the finished agarose gels were photographed (Fluor S, Biorad) using the parameters in table 3.27.

Table 3.26: Components for 2-log DNA ladder (marker 0.1 – 10 kb)

Component	Source	Final concentration	Mass
2-log DNA ladder	NEB N3200S	25 ng·µl^{-1}	10 µl
5x DNA loading buffer	Table 3.25	1x	80 µl
Milli-Q Water	-	-	ad. 400 µl
Storage at 2 – 8 °C			

DNA fragments were excised under hand lamp LWUV light at λ = 366 nm or SWUV light at λ = 254 nm. These fragments are purified using method in chapter 3.7.7. Subsequently after DNA fragment excision, the gel was photo-documented.

Table 3.27: Parameters for photo documentation using Biorad's Fluor S and SYBR Safe 1x.

Parameter	Adjustment
Slit	2.8 – 4
Filter	Filter #1
Integration	Manual
Integration time	20 – 90 s
Light Source	EpiUV
Scan Width	80 mm or 160 mm
Resolution/Sensitivity	High sensitivity

3.9.3 Plasmid DNA Sequencing

Confirmation of sequence of the final expression vector or the purified PCR fragement was performed by single strand sequencing through GATC Biotech AG or MWG Eurofins. The sequencing reactions were performed according to the didesoxy method. Sequence alignments were performed using ClustalW2 (http://www.ebi.ac.uk/Tools/msa/clustalw2/).

3.9.4 Determination of glucose concentration

Glucose (α-D-(+)-glucose) concentrations in medium were measured using the EBIO compact glucose analyser (Eppendorf). The principle of glucose concentration determination is its enzymatic oxidation by O_2 and glucoseoxidase (GOD) resulting in gluconic acid as well as H_2O_2. Hydrogen superoxide is quantitatively measured by a subsequent platinum electrode.

Three standards and controls were applied respectively. The system was calibrated using a standardised 216.2 mg·dl^{-1} glucose solution in Milli-Q water (α-D-(+)-glucose monohydrate, Roth, 6887.1). This glucose solution was calibrated with an external standard (EKF Glucose

Standard 12 mmol·l^{-1}). Here, the in-house standard was 6-fold measured against the external standard and should be between 210.8 – 221.6 mg·dl^{-1} glucose (mean ± 2.5 %). In addition, solutions with 500.0 mg·dl^{-1} as well as 50.0 mg·dl^{-1} glucose were applied as controls. All solutions were sterile filtered (0.2 µm) and stored at ≤-15 °C.

For measurement, the system and dilution buffer (Table 3.28) was tempered to room temperature (20 – 30 °C) and filled within the appropriate basin in the glucose analyser. The standard (216.2 mg·dl^{-1}), the controls (216.2 mg·dl^{-1}, 500.0 mg·dl^{-1}, 50.0 mg·dl^{-1}) and all samples were mixed vigorously with 1000 ml system and dilution buffer. The added amount was always 40 µl for each.

The glucose concentrations were calculated by determination of the mean values and standard deviation of theoretical and biological replicates (Table 3.10).

Table 3.28: Components of the system and dilution buffer for glucose determination

Component	Source	Final concentration
Na$_2$HPO$_4$	Roth P030.2	5.0 g·l^{-1}
NaH$_2$PO$_4$·H$_2$O	Roth 2370.1	1.3 g·l^{-1}
Sodium benzoat	Roth 8548.1	0.3 g·l^{-1}
EDTA(·2Na·2H$_2$O)	Roth 8043.3	0.5 g·l^{-1}
Sodium chloride	Roth 3957.1	5.1 g·l^{-1}

Solved in Milli-Q Water. The resulting pH should be 7.3 ± 0.1 automatically. Solution is sterile filtered 0.2 µm. Storage at 2 – 8 °C. Before measurement, the solution should have reached 20 – 30 °C.

3.9.5 hIgG ELISA

The concentrations of hIgG within culture medium were determined by enzyme-linked immune sorbent assay. Here, the constant hIgG Fcγ region is detected.Therefore, capture antibody (AffiniPure F(ab')$_2$ rabbit anti-human (Fcγ fragment specific, Jackson Immuno Research, 309-006-008) was diluted to 5 µg·ml^{-1} in PBS 1x (BioChrom, L182-50) and coated on MaxiSorp 96-well plates (Nunc, 442404) for 16 – 18 hours at 2 – 8 °C using 100 µl each well. In following, the uncoated regions were blocked with a 3.0 % albumin fraction V solution for three hours at 20 – 30 °C using 200 µl each well. In parallel to blocking incubation, the standards and controls were prepared and diluted in fresh sterile culture medium (Table 3.29). In addition, the samples (culture supernatants, chapter 3.2.1) were diluted 1:2500 – 1:10000 in fresh culture medium in respective to their putative hIgG concentration. After blocking, the wells were washed with 200 µl·well^{-1} PBS 1x once and subsequently 2x 100 µl of each standard, control as well as sample were added as technical duplicates. After incubation for one hour at

20 – 30 °C, the wells were washed with 4x 200 µl PBS 1x/0.5 % Tween-20 (Roth, 9127.1) (PBS-T) and with 200 µl PBS 1x once. Subsequently, 100 µl·well^{-1} of the detection antibody solution (1:5000, POD-conjugated AffiniPure F(ab')$_2$ fragment goat anti-human IgG (Fcγ fragment specific), Jackson I*mmu*no Research, 109-036-098) was added and incubated for one hour at 20 – 30 °C. After incubation, the wells were washed with 4x 200 µl PBS-T and with 200 µl PBS 1x once. Subsequently, 100 µl·well^{-1} substrate solution (Table 3.30) was added and incubated for 10 – 15 minutes at 20 – 30 °C.

Table 3.29: Standards and controls for hIgG Fcγ ELISA.

Standards		Controls	
Component	**Source**	**Component**	**Source**
ATROSAB	Celonic F10179	hIgG, purified I*mmu*noglobulin	Sigma I2511
Code	**Concentration**	**Code**	**Concentration**
S1	250 ng·ml^{-1}	K1	10 ng·ml^{-1}
S2	125 ng·ml^{-1}	K2	50 ng·ml^{-1}
S3	62.50 ng·ml^{-1}	K3	100 ng·ml^{-1}
S4	31.25 ng·ml^{-1}	K4	200 ng·ml^{-1}
S5	15.63 ng·ml^{-1}	**Blank**	
S6	7.81 ng·ml^{-1}	**Code**	**Concentration**
S7	3.91 ng·ml^{-1}	B	0.00 ng·ml^{-1}

Table 3.30: Substrate solution for POD-conjugated secondary antibodies.

Component	Source	Final concentration
0.1M K$_2$HPO$_4$	Roth P749.1	25 mM
0.1 M Citric acid	Roth 6490.3	25 mM
OPD (o-phenylene diamide) tablette	Sigma P5412	1x / 50 ml
30 % H$_2$O$_2$	Roth 8070.2	20 µl / 50 ml
All components were solved in given order in 25 ml Milli-Q water at 20 – 30 °C		

The reaction was stopped by adding 50 µl·well^{-1} 3M HCl (Roth, 6331.1) and quickly measured at λ = 490 nm/690 nm (wave corrective due to well plate irregularity). The resulting absorbance and recalculated concentrations were used for further calculation (Table 3.31).

3.9.6 SDS-PAGE

For qualitative or semi-quantitative hIgG product analysis SDS polyamide gel electrophoresis (SDS-PAGE) was performed. Here, the undiluted or diluted supernatant (13 µl) was mixed with 5 µl NuPAGE LDS Sample Buffer 4x (Invitrogen, NP0007) as well as 2 µl DTT 10x (0.5 M) (DL-1,4-dithiothreitole: Roth, 6908.3) and incubated for five minutes at 95 °C. After cooling, the denaturated samples were fully loaded on a NuPAGE® Bis-Tris 4-12 % Gel 1.0 mm, 12 well (Life Technologies, NP0322BOX) in 1x MOPS buffer (Table 3.32).

In addition, 10 µl Precision Plus Protein™ All Blue (BioRad, 161-0373) was added to each gel for size prediction. Finally, the set up was constantly run for 50 minutes at 200 V (Start: 100 – 115 mA; end: 60 – 70 mA each gel). The polyamide gel was carefully unhinged and washed three times with Milli-Q water for 5 – 10 minutes at 20 – 30 °C and 50 rpm (Amplitude: 20 mm). Here, the gel should be fully covered with Milli-Q water. For staining, the gel was incubated with SimplyBlue™ SafeStain (Invitrogen, LC6060) for one hour and the background was subsequently decolourised with 3x – 5x Milli-Q water incubation for at least 30 minutes each step. All incubation steps were performed at 20 – 30 °C as well as 50 rpm (Amplitude: 20 mm). For documentation, the gels were recorded and stored by a common colour scanner with the highest possible dpi value and with RPG colours.

Table 3.31: Formulae for determination of cell culture dependent parameters

Mean product concentration / product titer †		
$$\bar{c} = \sum_{b,t=1}^{n} c_{b,t}$$	b:	number of biological replicates
	t:	number of technical replicates
	$c_{b,t}$:	individual product concentration
Standard deviation of product concentration †		
$$S_c = \sqrt{\frac{\sum_{b,t=1}^{n}\left(c_{b,t} - \bar{c}\right)}{n-1}}$$	n:	sample size (biological and technical replicates)
	S_c:	Standard deviation of the product concentration regarding biological and technical replicates
Cell specific productivity		
$$csp = \frac{\bar{c}}{ivcd} \cdot 10^6$$	ivcd:	integral viable cell density
	csp:	cell specific productivity [pg·cell^{-1}·d^{-1}]
Volumetric productivity		
$$vp = \frac{\bar{c}}{t_c}$$	t_c:	culture duration to reach respective product concentration
	vp:	volumetric productivity [g·l^{-1}·d^{-1}]

† applicable for other parameters

Table 3.32: Components for MOPS buffer 20x

Component	Source	Final concentration
MOPS ·Na	Roth 6927.2	1.0 M
Tris	Roth 5429.1	1.0 M
SDS	Roth 2326.1	70.0 mM
EDTA(·2Na·2H$_2$O)	Roth 8043.3	20.0 mM

All components were solved in 500 ml Milli-Q water at 20 – 30 °C and stored at 2 – 8 °C in dark.
Prior to use, the solution was freshly diluted in Milli-Q water resulting 1x MOPS buffer.

3.10 Data compilation and computation

All raw data were manually documented on data sheets and transferred to Excel sheets for computerised documentation and calculation. Data of ELISA measurements were calculated using SoftmaxPro software 8.1 (Molecular Devices). All data were also transferred to Microsoft Office Excel 2007 for calculation of the final concentrations regarding the performed dilutions. All raw as well as calculated electronic data are safely archived or stored at proper server systems to achieve highest safety.

Protein and DNA/RNA sequences were aligned using ClustalW2 or ClustalΩ using the described recommendations (http://www.ebi.ac.uk/services). The consensus symbols at output alignments describe that the amino acid or nucleotide within a column is identical in all sequences in the alignment (*) (homologous), that conserved substitutions (:) or semi-conserved substitutions have been observed (.).

The phylogenetic distance matrices were generated using ClustalW2 Phylogeny (http://www.ebi.ac.uk/services) by application of the application file from ClustalW2 alignment and the option distance matrix for tree format.

Other online prediction tools used are referred directly.

3.11 Material

Following, the used equipment and devices as well as consumables are shown in tables 3.33 and 3.34, respectively.

Table 3.33: Equipment and devices.

Equipment / Device	Name	Supplier
Digital Imaging System	Fluor S	Biorad
Gel electrophoresis chamber	Easy Cast	Owi Scientific
Power Supply	E865	Consort
Heat Block	Thermomix Comfort	Eppendorf
Benchtop Centrifuge	Biofuge fresco & pico	Heraeus/Kendro
Centrifuge	Z383K	Hermle
Centrifuge	Multifuge 3S-R	Heraeus
Water Bath	1002	GFL
Autoclave	HST-4X5X6(8)-ES1073	Zirbus
Photometer	Ultrospec 100E	Pharmacia
Cleanbench	BDK-S1800	BDK
Pipettes	-	Gilson, Eppendorf and Rainin
Incubator	Incucell	MMM Group
Incubator Shaker	Infors HT	Infors
UV lamp	NU-4-KI	Konrad Benda
CO2 incubator	Heracell	Heraeus
Temperatur incubator Spinner	SI-01	GFL
Orbital shaker (Ampl.: 20 mm)	Rotamax 120	Heidolph
Orbital shaker (Ampl.: 30 mm)	Ovan Midi	Ovan, Spain
Orbital shaker (Ampl.: 30 mm)	SK-300	JeioTech, Japan
Erlenmeyer flask 100 ml	DURAN®	Schott
Erlenmeyer flask 500 ml	DURAN®	Schott
Erlenmeyer flask 1000 ml	DURAN®	Schott
Clean bench	KR-130 BW	Kojair
Microscope (invers)		Nikon
Counting chamber	Neubauer, improved	Marienfeld
Waterbath		DSL/DASGIP
Pipetboy		Costar
Nucleofector II		Amaxa/Lonza
Cryocontainer	Mr. Frosty	Nalgene
Liquid chromatography	Äkta explorer	Amersham pharmacia biotech
Water supply	Milli-Q biocel	Millipore

Table 3.34: Consumables

Name	Supplier	Catalogue Number
Reaction tubes 1.5ml, for RNA/DNA storage	Sarstedt	72.706.200
Reaction tubes 1.5ml	Nerbe plus	04-210-1100
Reaction tubes 2.0ml	Nerbe plus	04-230-1100
Filter tips 1000µl	Peqlab	81-1050
Filter tips 200µl	Peqlab	81-1040
Filter tips 10µl	Peqlab	81-1031
Stripettes 50ml	Corning	4490
Stripettes 25ml	Corning	4489
Stripettes 10ml	Corning	4488
Stripettes 5ml	Corning	4487
Tubes 50ml	Nerbe plus	02-572-7001
Tubes 15 ml	TPP	91015
PMMA disposable cuvette	Roth	P951
UV cuvettes micro	Brand	759220
disposable scalpel	Braun	5518040
T-25	TPP	90026
T-75	TPP	90076
6-well plates	TPP	92006
12-well plates	TPP	92012
24-well plates	TPP	92024
96-well plates	TPP	92696
96 well plates, MaxiSorp (ELISA)	Nunc	442404
200 µl filter tips (ELISA)	Rainin	RT-L200F
200 µl filter tips	Nerbe plus	81-1040
1,5 ml reaction tubes	Sarstedt	72.690.001
2 ml reaction tubes	Nerbe plus	04.230.1100
2 ml cryo tubes	TPP	89020
Sterilefilter, Acrodisc 25mm	Pall	4192
Bottle top filters 150ml	Millipore	SCGPT01RE
Bottle top filters 500ml	Millipore	SCGPT05RE

3.12 Terms and acronyms

Species-specific italic acronyms (e.g. *cgr, mmu, rno* or *hsa*) are further used to clearly depict the respective homologous biomolecule of a species (*C. griseus, M. musculus, R. norvegicus* or *H. sapiens*). For discussion (Chapter 5), the respective published acronyms as well as terms are used. Rodent biomolecule acronyms are shown in an initial capital letter followed by lower case letters (e.g. Hsp70), whereat human acronyms are throughout shown in capital letters.

3.13 Cell line terminology

All cell lines derived from one cryopreserved stock CHO-K1 (ATCC, CCL-61, Lot 3442943) (Cell line K20-3 derived from another cryo differing in lot number). In Figure 3.1, the origin of all cell lines developed in this thesis is shown.

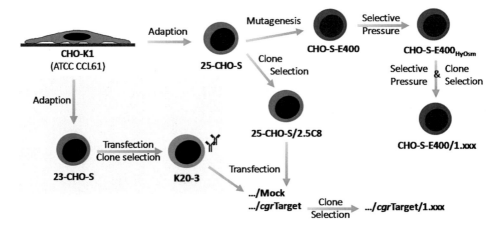

Figure 3.1: Table of origin of all cell lines developed and used in this thesis. The establishment of 23-CHO-S and the resulting antibody producing clone K20-3 was not performed within this thesis (shaded). The suffix "-S" states for suapension and is not connected with other commercial products or cell lines. The suffix "-E400" implements the treatment with 400 µg·ml⁻¹ Ethylmethanesulfonate (EMS). The note "HyOsm" describes the applied selection pressure at high osmolalities and is not used for subsequently derived clones furthermore. The cell line clones derived from Limited Dilution (LD) and functional selection are described as following: #LD."plate&coordinates" (e.g. 2.5C8) or #LD."consecutive numbering" (e.g. 1.104) (solely used for CHO-S-E400$_{HyOsm}$ derived clones).

4   Results

In the following chapters, all steps and final results are presented in detail. The final discussion is performed in chapter 5. For better understanding and displaying the strategy behind the steps, some hypothesises as well as argumentations are implemented within the presentation of the results. Herein, the necessary generation of a clonal CHO suspension cell line is described (Chapter 4.1) with subsequent identification by differential expression, verification and validation by stable transfection of high cell density (HCD) factors.

4.1   Generation of a clonal CHO cell line with capability growing at HCD

This section focuses in the generation process of a clonal CHO suspension cell line that is able to grow at high cell densities (HCD) beginning from adherent CHO-K1.

4.1.1   Adaption of CHO-K1 cells to CM1035

The adherent CHO-K1 from ATCC (CCL-61, Lot 3442943) at passage 4 in Kaighn's F12 + 10 % FCS was directly adapted to CM1035 medium supplemented with 4 mM L-glutamine (Gln), cultivated in spinner flasks for further eleven passages and cryopreserved. In parallel, at passage 16 a batch culture was performed to elucidate the performance of the new 25-CHO-S pool cell line (Figure 4.1).

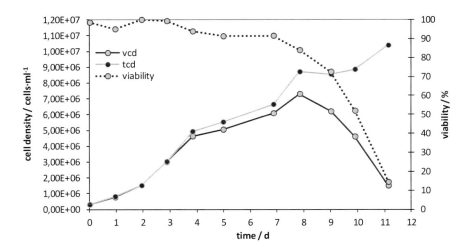

Figure 4.1:   Batch culture of 25-CHO-S at passage 16 in CM1035 + 4 mM Gln. The maximal viable cell density (vcd) during this representative batch culture was $7.3 \cdot 10^6$ cells·ml^{-1} at a viability of 84 %. The doubling time in exponential phase was 22 hours. Culture conditions: 37 °C, 80 rpm, 2.5 sL·h^{-1}, 21 % O_2, 5 % CO_2, 250 ml spinner flask, 80 ml cell suspension (n = 1, tcd: total cell density).

4.1.2 Chemical mutagenesis and mutated cell regeneration

Mutagenesis of cell line 25-CHO-S was performed using 400 µg·ml^{-1} of the alkylating agent ethylmethane sulphonate (EMS) in CM1035 + 4 mM Gln in suspension culture (Figure 4.2).

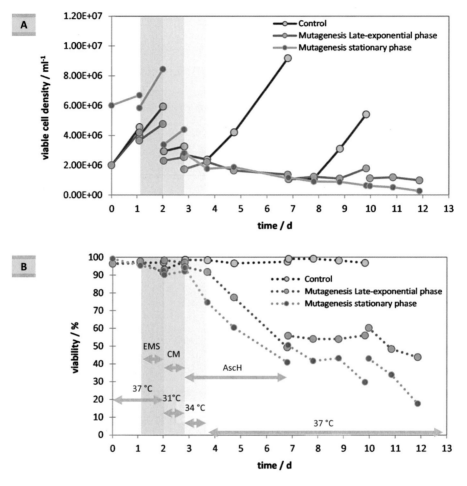

Figure 4.2:   Mutagenesis of 25-CHO-S using 400 µg·ml^{-1} ethylmethane sulfonate (EMS). Mutagenesis at late-exponential phase was performed with cells possessed high growth rate at a low start cell density. Mutagenesis at stationary phase was performed with slowly growing cells at high starting cell density. Control cell culture was performed identically to modus of mutagenesis at late-exponential phase without EMS addition. For regeneration, temperature shifts were applied as well as conditioned medium (CM) and 5.0 mg·ml^{-1} ascorbic acid (AscH) was consecutively added. The course of viable cell density (A) and viability (B) are shown above. Culture conditions: CM1035 + 4 mM, 37 °C, 80 – 105 rpm, 2.5 – 3.0 sL·h^{-1}, 21 % O_2, 5 % CO_2, 250 – 1000 ml spinner flask, 80 – 200 ml cell suspension (n = 1).

The cells were seeded on day zero in 70 ml CM1035 + 4 mM Gln at cell densities of 2.0·10^6 cells·ml^{-1} for control and mutagenesis at late-exponential phase (MLEP) and

$2.0 \cdot 10^6$ cells$\cdot$ml^{-1} for mutagenesis at stationary phase (MSP). Each growth phase should possess different cellular targets for EMS-mediated disruption.

After one day, 3200 µg$\cdot$ml^{-1} EMS in 10 ml CM1035 + 4 mM Gln were added to reach the final concentration of 400 µg$\cdot$ml^{-1} and incubated for 22 hours (37 °C, 80 rpm, 2.5 sL$\cdot$h^{-1}, 21 % O_2, 5 % CO_2, 250 ml spinner flask).

Following the cells were recovered by centrifugation, washing with PBS 1x and resuspension in conditioned medium (150 ml) from exponentially growing 25-CHO-S cell cultures to provide additional growth factors (Further cultivation: 105 rpm, 3.0 sL$\cdot$h^{-1}, 21 % O_2, 5 % CO_2, 1000 ml spinner flask). The culture temperature was shifted to 31 °C to increase expression of anti-apoptotic proteins (Slikker *et al.* 2001; Fu *et al.* 2004). Therefore, these manipulations should help the cells to recover properly.

One day after, 50 ml CM1035 + 4 mM Gln + 5.0 mg$\cdot$ml^{-1} ascorbic acid (AscH) were added to provide partial fresh medium and overcome possible oxidative stresses and apoptosis (Witenberg *et al.* 1999). To ensure growth, temperature was gradually increased to 34 °C for one day and finally to 37 °C.

After 75 hours in culture, the cells were recovered by standard centrifugation (Chapter 3.2) and resuspended in 50 ml of the supernatant, which was prior centrifuged (2600$\cdot$g, five minutes, 20 – 30 °C) to remove the cell debris. Additional 150 ml CM1035 + 4 mM Gln was added. After further three days in culture, the cells were recovered again but by mild centrifugation (60$\cdot$g, five minutes, 20 – 30 °C) to enrich the viable cells (Takagi *et al.* 2000). The cells were resuspended in the respective supernatant, which was prior centrifuged again (2600$\cdot$g, five minutes, 20 – 30 °C). After another two days the cells were cryopreserved (Chapter 3.4) for saving the cells.

Here, the treatment with EMS led to massive growth inhibition and consistent loss on viability compared to the control cell culture (Figure 4.2). Furthermore, EMS treatment of cells at stationary phase led to a more distinct reduction in cell number and viability than EMS treatment of cells at late-exponential phase. After thawing, this trend was continued leading to total abolishment of viability and cell number of 25-CHO-S cells EMS-treated at stationary phase.

In parallel, the viability of CHO-S-E400 cells (25-CHO-S cells, which were treated with 400 µg$\cdot$ml^{-1} EMS at late-exponential phase) quickly dropped after thawing (Figure 4.3). Therefore, the cells were centrifuged to add fresh medium (CM1035 + 4 mM Gln, 70 ml) without cell debris and putative pro-apoptotic factors. In the same manner, medium was continuingly changed at day 8 (70 ml medium) and 16 (50 ml medium) after thawing (37 °C, 105 rpm, 2.5 sL$\cdot$h^{-1}, 21 % O_2, 5 % CO_2, 250 ml spinner flask).

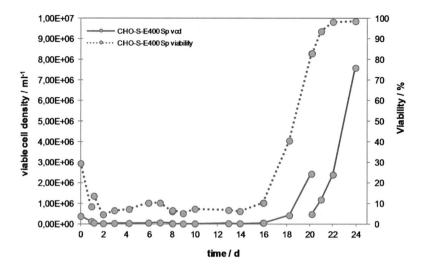

Figure 4.3: Regeneration of EMS-treated 25-CHO-S cell line CHO-S-E400 (former mutagenesis at late-exponential phase). After drop in viability, the cells were kept in culture. After 16 days the cells started to grow again. Culture conditions: CM1035 + 4 mM, 37 °C, 105 rpm, 2.5-3.0 sL·h^{-1}, 21 % O_2, 5 % CO_2, 250 – 1000 ml spinner flask, 50 – 200 ml cell suspension (n = 1).

At day 16, a slight increase in viability was observed, which resulted in total recovery of the cells in following days. At day 20 after thawing, the total cell number in 40 ml cell suspension was transferred to a bigger spinner volume of 200 ml by adding CM1035 + 4 mM Gln (37 °C, 105 rpm, 3.0 sL·h^{-1}, 21 % O_2, 5 % CO_2, 1000 ml spinner flask). After further four days in culture the cells were cryopreserved at passage 33, viability of 98.7 % and with a mean growth rate of 22.9 hours (Figure 4.3).

Compared to its progenitor cell line 25-CHO-S, CHO-S-E400 reached slightly higher cell densities (Figure 4.4). During these repetitive batches in CM1035 + 4 mM the viability course slightly downgraded compared to 25-CHO-S viability pattern.

In batch culture, freshly thawed CHO-S-E400 showed higher growth rates as well as an increased maximal cell density than parallel cultured 25-CHO-S cells (Figure 4.5). The viability curves of both cell lines were similar, whereas CHO-S-E400 faded before 25-CHO-S possibly by earlier nutrient deprivation due to higher integral viable cell density (1.6-fold, >80 % viability).

Nevertheless, the increase in cell density was not enough to be significant for further differential expression studies. Therefore, higher selection pressures were set (Chapter 4.1.3).

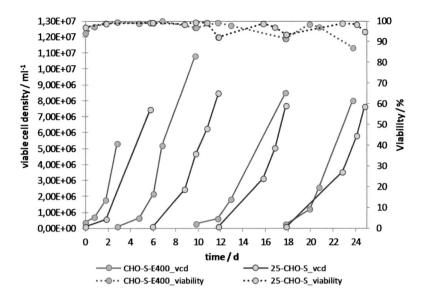

Figure 4.4: Growth characteristic of CHO-S-E400 cell line compared to the progenitor cell line 25-CHO-S. The cell line stably grew in CM1035 + 4mM, but showed fading viability performance. Culture conditions: 37 °C, 105 rpm, 2.5 sL·h^{-1}, 21 % O$_2$, 5 % CO$_2$, 250 ml spinner flask, 50-70 ml cell suspension (n = 1).

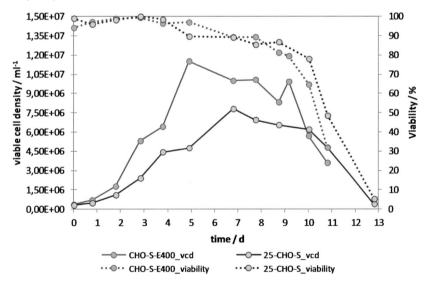

Figure 4.5: Batch comparison of CHO-S-E400 to 25-CHO-S in CM1035 + 4 mM Gln. Freshly thawed CHO-S-E400 showed higher growth rate and maximal cell density than 25-CHO-S (P19). Culture conditions: 37 °C, 105 rpm, 2.5 sL·h^{-1}, 21 % O$_2$, 5 % CO$_2$, 500 ml spinner flask, 100 ml cell suspension (n = 1).

Finally, the fragmentation of the nuclei was determined by formaldehyde fixation and DAPI staining (Chapter 3.9.1). Therefore, samples of the previous batch (Figure 4.5) were taken at day 3 and were fixed as well as stained (Figure 4.6).

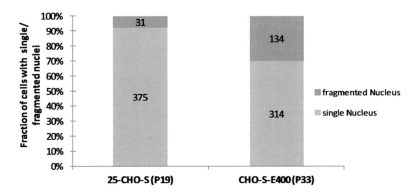

Figure 4.6: Fragmentation of nuclei in exponentially growing CHO-S-E400 (P33) and 25-CHO-S (P19). The respective amounts represent the counted nuclei. (One nucleus per cell, $n_{biological} = 1$)

Here, in CHO-S-E400 the grade of nucleus fragmentation was 3.9-fold higher compared to its progenitor cell line 25-CHO-S. The percentage of fragmented nuclei in exponentially growing CHO-S-E400 culture was 30 % compared to 8 % for 25-CHO-S, which indicated successful mutation and chromosomal rearrangements.

4.1.3 Selection of CHO-S-E400 towards growth at high osmolality and glutamine absence

Based on the mean doubling time in figure 4.5, which was calculated to be 23.55 hours until day 5, a theoretical back calculation was performed to estimate the initial cell number. For the cell number estimation the putative growth start was set to day 8 in figure 4.3 until day 20. Therefore, based on $1.22 \cdot 10^8$ cells at day 20, the cell number was back calculated to day 8 resulting an estimated putative cell number of $2.4 \cdot 10^4$ cells.

This huge number of cells needed to be reduced for proper single cell cloning. Following the strategies in chapter 2.3, several passages at high osmolality (525 – 600 mOsmol·kg^{-1}) and absence of or reduced L-glutamine (0 – 1 mM) were performed (Figure 4.7) to generate resistant cell population against high osmolality enabling higher cell densities. The increased osmolality was accomplished by final addition of sodium chloride (100 mM or 137.5 mM, respectively) on the basis that CM1035 had an osmolality of 325 – 350 mOsmol·kg^{-1} (Certificate of analysis, HyClone). In addition, the ongoing culture was stopped and passaged after viability declined below 75 % to enrich more resistant against nutrient deprivation, apoptosis as well as osmolality. Further, it was assumed that due to high evaporation rate in non-humidified spinner flask (Table 3.4) and lactate formation osmolality successively increased (approx. $1 - 2\ \%\cdot d^{-1}$) and contemporaneously the selection pressure increased.

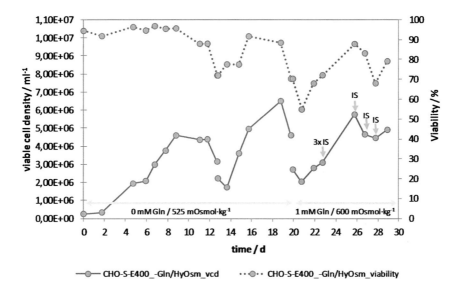

Figure 4.7: Selektion CHO-S-E400 at high osmolality ($525 - 600\ mOsmol\cdot kg^{-1}$) and glutamine absence. Freshly thawed CHO-S-E400 were directly cultured in CM1035 + 100 mM NaCl for two passages and CM1035 + 137.5 mM NaCl + 1 mM Gln for one passage. IS indicates feeding of IS CHO Feed-CD ($2\%\cdot d_{equivalent}^{-1}$). Culture conditions: 37 °C, 105 rpm, 3.0 sL·h^{-1}, 21 % O$_2$, 5 % CO$_2$, 1000 ml spinner flask, 180 ml cell suspension (n = 1).

The CHO-S-E400 cells were freshly thawed to maintain the original clone consistency. The first two passages were performed in CM1035 + 100 mM NaCl ($525\ mOsmol\cdot kg^{-1}$). Here, the growth was decelerated and cell densities stayed beyond $7.0\cdot 10^6$ cells·ml^{-1}. Anyway, the batch culture was performed until viability decreased beyond 75 % and the cells were centrifuged by default (Chapter 3.2) and resuspended in fresh medium. Here, less the cells for cryopreservation all cells were introduced into the new culture to additionally set a selection pressure towards high cell density. Nevertheless, possibly due to apoptotic program onset the cell number initially decreased in the first day of culture.

During the third and last selection step before limited dilution as well as single cell cloning, the cells were resuspended in CM1035 + 1 mM Gln + 137.5 mM NaCl ($600\ mOsmol\cdot kg^{-1}$) to increase the initial nutrition status and osmolality. To overcome the negative effects of increased osmolality, the culture was fed with $2\%\cdot d_{equivalent}^{-1}$ IS CHO Feed-CD XP ($280 - 320\ mOsmol\cdot kg^{-1}$). This feeding was suitable to overcome the increase in osmolality by evaporation and additionally support the cells with additional nutrients. Furthermore, the growth rate decreased and after nine days in culture, the selection was stopped for initiating single cell cloning by Limited Dilution (Chapter 4.1.4). Before, cells were harvested during the second passage (day 15, Figure 4.7) to characterise the growth in CM1035 + 4mM Gln under batch and non-optimised fedbatch conditions (Figure 4.8).

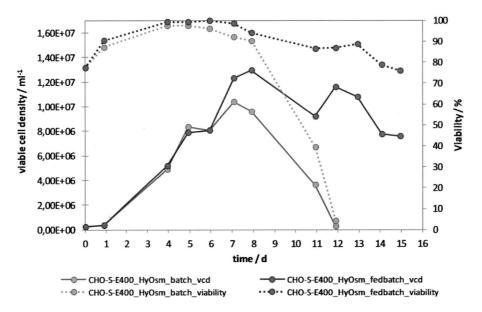

Figure 4.8: Batch and fedbatch culture of CHO-S-E400 prior selected at high osmolality. Feed was performed in fedbatch culture at day 1 (4 ml) and then 2 ml·d$_{equivalent}^{-1}$ IS CHO Feed-CD XP. Culture conditions: CM1035 + 4 mM Gln, 37 °C, 105 rpm, 2.5 sL·h^{-1}, 21 % O$_2$, 5 % CO$_2$, 500 ml spinner flask, 90 ml cell suspension (n = 1).

The batch culture of CHO-S-E400$_{HyOsm}$ showed high similarities to CHO-S-E400 batch culture (Figure 4.5). Therefore, the selection had at least no adverse effects. On the other hand, the cells within the fedbatch culture reached higher cell densities at identical growth rate. In addition, feed provoked higher viabilities and prolonged viable cell culture at high cell densities. It seems, in batch culture the cells rapidly underwent autophagy/necrosis due to nutrient deprivation. Therefore, the mutated CHO-S-E400 and CHO-S-E400$_{HyOsm}$ carry the potential for growing at high cell densities.

4.1.4 Single cell cloning of CHO-S-E400$_{HyOsm}$

Single cell cloning of CHO-S-E400$_{HyOsm}$ was performed as shown in table 3.22. Cells of third selection passage on day 29 (Figure 4.7) were directly diluted in prewarmed 90% CM1035 + 4 mM Gln / 10 % IS CHO Feed-CD XP (30 – 37 °C) to a final concentration of 10 cells·ml^{-1}. This cell suspension was portioned on twelve 96-well culture plates with 100 μl·well^{-1} theoretically resulting 1 cell·well^{-1}. Therefore, the total cell number seeded was 1152 cells. In 67.3 % of the wells single cells were found (775 cells). After six days cells were visibly grown in 500 wells and the first 48 clones with ≥60 % confluency could be transferred into 6-well plates with 2 ml

CM1035 + 4 mM Gln each, followed by 72 clones after further two days and 36 clones after further five days. Here, the fast growing cells were preferred due to the putative beneficial growth characteristic during further selection steps.

The resulting 156 CHO-S-E400 clones (E_{LD}: 20.1 %; Table 3.23) were continuingly cultured applying 33 – 36 passages of 1:10 – 1:20 dilution each 7 – 12 days resulting 2 ml cell suspension CM1035 + 4 mM Gln. This step was applied for stabilising the phenotype.

4.1.5 Selection of CHO-S-E400$_{HyOsm}$ clones

After extent continuous culture in static 6-well plates, each cell suspension (2 ml) was centrifuged, resuspended in 10 ml CM1035 + 4 mM Gln and transferred to 50 ml bioreactor filter tubes (BFT50). Subsequently, the first clone selection was performed in batch cultivation for 14 days selecting clones suitable for long-term batch cultivation, efficient nutrient consumption and resistance against apoptosis (37 °C, 200 rpm, amplitude: 30 mm, 5 % CO_2, aeration: A-D).

The qualitative parameters for clone refusal after 14 days were medium staining (viable: amber, yellow; Dead: cream, milky yellowish-white) and grade of aggregation, whilst low aggregation was preferred and cell cultures with total aggregation considered as dead. Two clones did not grow under these culture conditions. Subsequently, 48 viable cell cultures were centrifuged, resuspended in 5 – 10 ml CM1035 + 4 mM Gln and cultured for 8 – 12 days (37 °C, 200 rpm, amplitude: 30 mm, 5 % CO_2, aeration: A-D). To these high cell density cultures up to 20% IS CHO Feed-CD XP was fed. The same qualitative parameters for clone refusal were applied, while the cell density of each low-aggregated and viable cell culture was manually counted (22 clones). Finally, five clones were viable enough for further cultivation. The growth characteristics of the first three clones are comparatively shown in figure 4.9. Here, the clone CHO-S-E400/1.056 revealed the highest cell density, highest growth rate and best viability course between the compared three clones. In Figure 4.10, clone CHO-S-E400/1.056 was then compared with the two residual clones. The comparison of the CHO-S-E400 clones 1.056, 1.104 and 1.150 revealed that CHO-S-E400/1.104 possessed the highest maximal cell density as well as growth rate in batch as well as primitive fedbatch culture. The viability course was slightly higher. In addition and contrast to the other clones, CHO-S-E400/1.104 did not aggregate and the cells were significantly smaller ($\varnothing_{cell} \approx 10 – 12$ μm, compared to 14 μm). Therefore, the clone CHO-S-E400/1.104 was used for the transcriptomic studies (Chapter 4.2).

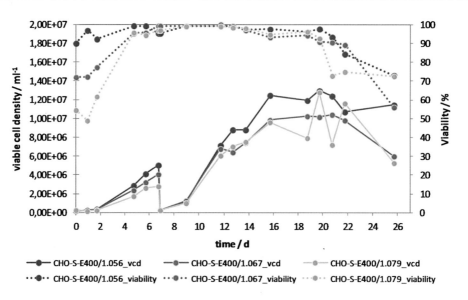

Figure 4.9: Comparison of CHO-S-E400 clones 1.056, 1.067 and 1.079. Feed was performed in fedbatch culture with IS CHO Feed-CD XP at day 5 (2 ml), 9 (1 ml), 13 (1 ml), 20 (1 ml), 22 (2 ml). Culture conditions: CM1035 + 4 mM Gln, 37 °C, 200 rpm, amplitude: 30 mm, 5 % CO_2, aeration: A-D, BFT50, 10 ml cell suspension (n = 1).

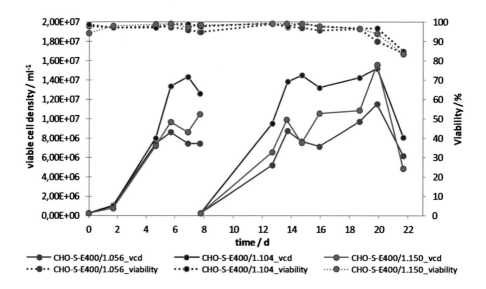

Figure 4.10: Comparison of CHO-S-E400 clones 1.056, 1.104 and 1.150. Feed was performed in fedbatch culture with IS CHO Feed-CD XP at day 16 (1.5 ml) and 20 (2 ml). Culture conditions: CM1035 + 4 mM Gln, 37 °C, 200 rpm, amplitude: 30 mm, 5 % CO_2, aeration: A-D, BFT50, 10 ml cell suspension (n = 1).

4.1.6 Single cell cloning of 25-CHO-S

Single cell cloning of 25-CHO-S was performed as shown at table 3.22. The first limited dilution was performed without single cell cloning in CM1035 + 4mM Gln and 75 % CM1035 + 4mM Gln / 25 % PBS 1x, two plates each as well as 1 cell·well^{-1}. Here, twelve subclones were recovered and three subclones finally selected by the highest reached cell density, low aggregation as well as best viable appearance.

Subsequently, these three clones were separated by single cell cloning, at which each clone was seeded in CM1035 + 4 mM Gln on two plates and 0.5 cell·well^{-1}. Total single cell number was 50 cells, from which five clones were recovered, four clones survived in T-75 culture (E_{LD}: 8.0 %; Table 3.23) and finally were cultured in shaken BFT50 culture (Figure 4.11).

Here, the four 25-CHO-S clones are compared and their ability for growth in shaken culture system is observed. The 25-CHO-S clone 2.2F3 was not able to grow in shaken cultures under given condition. Interestingly, both clones 2.4B6 and 2.4F3 grew fast, but suddenly died after short culture periods. In addition, both clones highly aggregated in shaken culture.

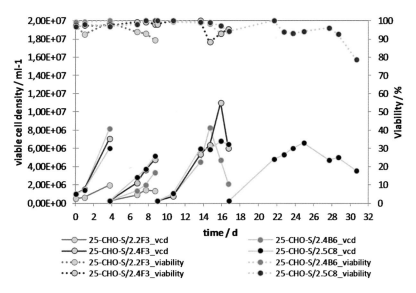

Figure 4.11: Comparison of 25-CHO-S clones 2.2F3, 2.4B6, 2.4F3 and 2.5C8. Feed was performed in fedbatch culture with IS CHO Feed-CD XP at day 16 (1.5 ml) and 20 (2 ml). Culture conditions: CM1035 + 4 mM Gln, 37 °C, 200 rpm, amplitude: 30 mm, 5 % CO_2, aeration: A-D, BFT50, 10 ml cell suspension (n = 1).

Finally, 25-CHO-S/2.5C8 showed very low aggregation and constantly grew in shaken cultures reaching similar maximal cell densities (up to $6.6 \cdot 10^6$ cells·ml^{-1}). These cell densities were slightly lower than cell densities of 25-CHO-S pool cell line cultured in spinner flasks (10 % lower), which might be an effect of culture system. Nonetheless, the ability for cell aggregation, the growth rates as well as viability characteristics were comparable between clone 2.5C8 and pool 25-CHO-S cell line. Furthermore, 25-CHO-S/2.5C8 showed similar cell aggregation and viability characteristics as mutated clone CHO-S-E400/1.104 (Figure 4.12). In addition, the maximal cell density in reiterating batch culture (before adding feed) of CHO-S-E400/1.104 was 2.2-fold higher compared to 25-CHO-S/2.5C8. Therefore, the cell lines phenotypic differed in growth rate and maximal cell density, which was suitable for applying this pair in differential transcriptomics.

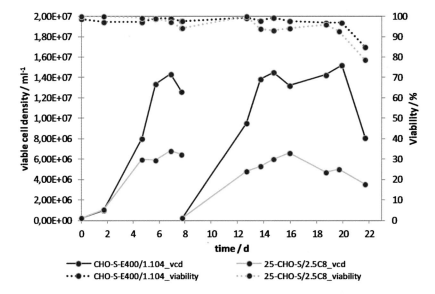

Figure 4.12: Comparison of CHO-S-E400/1.104 with 25-CHO-S/2.5C8. Feed was performed in fedbatch culture with IS CHO Feed-CD XP at day 16 (1.5 ml) and 20 (2 ml). Culture conditions: CM1035 + 4 mM Gln, 37 °C, 200 rpm, amplitude: 30 mm, 5 % CO_2, aeration: A-D, BFT50, 10 ml cell suspension (n = 1).

Further, to overcome possible nutrient limitation and increase the difference between both clones, the fedbatch process should be optimised before sampling (Chapter 4.1.9).

4.1.7 Transfection and productivity of CHO-S-E400/1.104 and 25-CHO-S/2.5C8

During sampling, a difference in cell pellet size of 4-fold (1.104/2.5C8) was observed using the same cell number. This observation could not be explained by reduced cell diameter in CHO-S-E400/1.104 cells of approx. 3 μm only. Due to the higher surface/volume ratio of smaller cells, it was suggested that the clonal mutant cell line would possess a high specific productivity too. Thus, the cell lines were transfected with a vector encoding a humanised IgG (hIgG$_A$). The resulting pool cell lines showed similar growth characteristics but productivities behaved inversely to reached cell density (Figure 4.13).

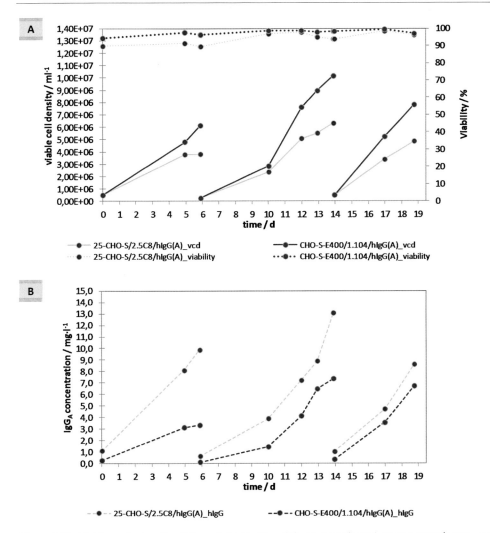

Figure 4.13: Viable cell density (A) and hIgG titer (B) of transfected CHO-S-E400/1.104 and 25-CHO-S/2.5C8 in repetitive batch culture. Medium: 87.5 % CM1035 / 10 % EfficientFeed A/B + 5 mM Gln. Culture conditions: 37 °C, 5 % CO_2, T-25, 5 – 6 ml cell suspension (n = 1).

The specific productivity of hIgGA-producing pool cell lines were 0.22 ± 0.11 pg·cell^{-1}·d^{-1} (CHO-S-E400/1.104 derived) and 0.56 ± 0.13 pg·cell^{-1}·d^{-1} (25-CHO-S/2.5C8 derived). Therefore, the mutagenesis by EMS presumably led to decreased production of extracellular matrix due to reduced pellet mass but also to impaired protein synthesis machinery. Consequently, the cell line CHO-S-E400/1.104 was not suitable for protein production even the productivity was not fully abolished. Anyway, because of the increased cell density the possibility to find a related factor was high enough to introduce the non-transfected cell lines within differential transcriptomics.

4.1.8 Accumulating cell culture CHO-S-E400/1.104 and 25-CHO-S/2.5C8

To determine the maximal possible cell density or cell density limit, CHO-S-E400/1.104 and 25-CHO-S/2.5C8 were cultured in serial batch cultures with applied centrifugation and cell number accumulation. Here, the cells were accumulated by using the same culture volume in a perfusion-like manner (Figure 4.14).

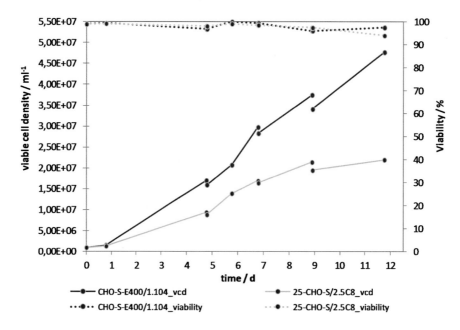

Figure 4.14: Determination of cell density limit. CHO-S-E400/1.104 and 25-CHO-S/2.5C8 were cultured in accumulating perfusion-like batch cultures using pre-fed media. Medium: 85 % CM1035 + 4 mM Gln / 15 % CHO CD EfficientFeed A/B (day 0-7); 70 % CM1035 + 4 mM Gln / 30 % CHO CD EfficientFeed A/B (day 7-12). Medium change was performed at day 5, 7 and 9. Culture conditions: 37 °C, 200 rpm, amplitude: 30 mm, 5 % CO_2, aeration: A-D (day 0-5); A-E (day 5-12), BFT50, 10 ml cell suspension (n = 1).

By this method, the maximal reached cell density for CHO-S-E400/1.104 was $4.77 \cdot 10^7$ cells·ml^{-1}, compared to $2.20 \cdot 10^7$ cells·ml^{-1} for 25-CHO-S/2.5C8. Therefore, the resulting ratio ($R_{1.104/2.5C8}$) was 2.2-fold and identical to batch culture above (Figure 4.12). Nevertheless, fedbatch culture was the method of choice for producing more reliable results without perturbation by centrifugation or osmotic pressure. Now, the fedbatch should be optimised (Chapter 4.1.9).

4.1.9 Optimised fedbatch cultures and sampling for factor identification

For differential transcriptomics, reliable samples with maximal spreading in cell density were needed, therefore the fedbatch cultivation process was roughly optimised and the reproducibility was determined. In addition, full batch or fedbatch cultures were avoided to keep the cell lines CHO-S-E400/1.104 and 25-CHO-S/2.5C8 in constantly viable state. First, the clonal cell lines CHO-S-E400/1.104 were 25-CHO-S/2.5C8 cultured in a limited batch (Figure 4.15, A) and fedbatch (Figure 4.15, B) cultures to possibly achieve samples at exponential as well as early stationary phase.

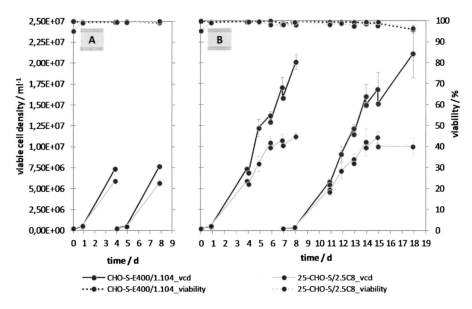

Figure 4.15: Comparison of CHO-S-E400/1.104 with 25-CHO-S/2.5C8 in repetitive fedbatch. To elucidate the reproducibility, two short-term batches (A) and fedbatches (B) were performed in parallel each. Second cultures were inoculated at day 4 or 7 using cells from the previous batch or fedbatch, respectively. Medium: CM1035 + 4 mM Gln. Feed (B): 625 µl 80 % IS CHO Feed-CD XP + 40 mM Gln) at day 4, 6, 7 or 11, 13, 14 respectively and 1.25 ml at day 15. Culture conditions: 37 °C, 200 rpm, amplitude: 30 mm, 5 % CO_2, aeration: A-D, BFT50, 10 ml cell suspension. Each point represents a mean of triplicates with the resulting standard deviation (n = 3).

Here, the growth was reproducible and similar cell densities were reached at each time point of culture and passage. Since cultures were set up in triplicates, the ratio of the significantly distinct (mean) maximal cell densities during the fedbatches ($R_{1.104/2.5C8}$) was 1.8- to 2.1-fold, respectively. In contrary to 25-CHO-S/2.5C8, the cell density of CHO-S-E400/1.104 increased after feed addition suggesting that the maximal possible cell density was still not reached. Due to these results, a more stringent feed regime was applied to further increase the maximal cell density in CHO-S-E400/1.104 fedbatch cultures (Figure 4.16).

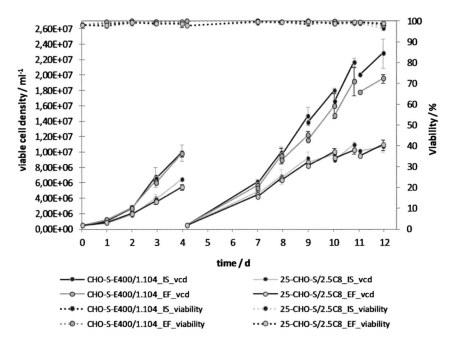

Figure 4.16: Sampling for transciptomics and qRT-PCR verification. CHO-S-E400/1.104 and 25-CHO-S/2.5C8 were cultured in repetitive fedbatches using two individual feed media in parallel. Sampling was performed at day 4 (exponential probes) and day 12 (stationary probes). Medium: 85 % CM1035 + 4 mM Gln / 15 % IS CHO Feed-CD XP or CHO CD EfficientFeed A/B. Feed IS: 80 % IS CHO Feed-CD XP + 40 mM Gln; Feed EF: 80 % CHO CD EfficientFeed A/B + 40 mM Gln. Feeding rate: 312.5 µl at day 2, 3 and 8; 625 µl at day 7 and 9; 937.5 µl at day 10 and 11. Culture conditions: 37 °C, 200 rpm, amplitude: 30 mm, 5 % CO_2, aeration: A-D, BFT50, 10 ml cell suspension. Each point represents a mean of quadruplicates with the resulting standard deviation (n = 4).

Based on already feed-supplemented medium (85 % CM1035 + 4 mM Gln / 15 % IS CHO Feed-CD XP) the feed ratio was increased by 25.6 % IS CHO Feed-CD XP (evaporation balanced) and total intake of 137.5 µmol L-glutamine (in total 4 mM + 12.6 mM Gln added). In parallel, CHO CD EfficientFeed A/B (50 % / 50 %) was applied instead of IS CHO Feed-CD XP, so feed-specific effects could be observed.

The experiment was performed in quadruplicates and showed again significant differences in cell density between CHO-S-E400/1.104 and 25-CHO-S/2.5C8 using both feed media individually, but viabilities for all samples were similar and over 96 %. In addition, application of CHO CD EfficientFeed A/B at CHO-S-E400/1.104 cultures led to lower cell densities than IS CHO Feed-CD XP, whereby this effect could not be observed for 25-CHO-S/2.5C8. The ratios (mean) maximal cell densities during the fedbatches ($R_{1.104/2.5C8}$) on day 12 were 1.8-fold for EF-fed and 2.1-fold for IS-fed cell cultures, respectively. On day 4 and within the exponential growth phase, the ratios were 1.8-fold for EF-fed and 1.5-fold for IS-fed cell cultures, respectively. The maximal cell densities for IS-fed CHO-S-E400/1.104 cell cultures were

$2.28 \cdot 10^7 \pm 0.19 \cdot 10^7$ cells·ml^{-1}, compared to $1.96 \cdot 10^7 \pm 0.05 \cdot 10^7$ cells·ml^{-1} in EF-fed CHO-S-E400/1.104 cell cultures. For cell cultures the growth duration, viability course and low aggregation ability were comparable.

Therefore, a limitation was suggested and due to this observation IS-fed cultures were excusively used for transcriptomics (Chapter 4.2.1), whereat samples of both feed media were used for quantitative RT-PCR verification (Chapter 4.2.2). In addition, the change in maximal cell density in IS-fed cultures was comparable with batch (Figure 4.12) and perfusion-like batch culture (Figure 4.14) and maximal possible difference in cell density was achieved.

4.2 Factor identification and verification

4.2.1 Transcriptomics by HiSeq 2000

As mentioned above, differential transcriptome analysis was performed using each three replicates of CHO-S-E400/1.104 and 25-CHO-S/2.5C8 IS-fedbatches at day 12 (Figure 4.16). The deeply frozen cell pellets ($2.5 \cdot 10^6$ cells, chapter 3.2.2) were processed and analysed by DNAVision S.A. in Belgium using Illumina's HiSeq 2000 sequencer. One lane per sample was used, therefore, $24.5 \cdot 10^6$ reads per sample to average were performed. The transcripts of each replicate and cell line were quantified and aligned to mouse genome database (ENSMUSG).

The raw data compromised 18265 genes within a range of $-12.40 \leq \log_2(FoldChange_{1.104/2.5C8})$ ≤ 9.22. For reducing the data volume, the p-value threshold was set to $y < 0.001$ and $\log_2(FoldChange_{1.104/2.5C8})$ to $-2.00 < x > 2.00$ resulting in 78 upregulated as well as 992 downregulated transcripts. In addition, the residual transcripts were manually screened for functional relevance, if applicable. Therefore, bulk and structural proteins as well as ribosomal fragments were screened out. Furthermore, functionally well-known proteins (e.g. chapter 2.5) were not considered for further analysis.

As a result, following transcripts were chosen for verification by quantitative RT-PCR (Table 4.1) and are discussed in chapter 5.2 and table 5.1. Subsequently, the resulting differently expressed transcripts were verified by quantitative RT-PCR (Chapters 4.2.2 and 3.6.3).

Table 4.1: HiSeq 2000 raw data of selected differentially transcribed factors

Transcript	Mean$_{1.104}$	Var$_{1.104}$ ‡	Mean$_{2.5C8}$	Var$_{2.5C8}$	FC$_{1.104/2.5C8}$ ‡	log2 (FC)	p-value
TPRp †	162,352	1,270	0,450	0,000	360,809	8,495	1,055E-03
9330101J02Rik	6,658	2,841	0,164	0,492	40,562	5,342	3,051E-07
Mir137	4,479	0,630	0,112	0,337	39,814	5,315	4,564E-05
AI661453	8,754	6,091	0,277	0,014	31,645	4,984	1,731E-04
Ankhd1	4,642	5,875	0,277	0,229	16,780	4,069	1,938E-04
Hmga1	1902,919	1,586	244,143	0,073	7,794	2,962	2,206E-04
BC068281	0,687	0,028	8,859	10,939	0,078	-3,689	9,472E-04
Ttc36 / HBP21	57,790	0,166	843,230	14,979	0,069	-3,867	4,587E-06
9330182L06Rik	0,281	0,035	4,715	11,397	0,060	-4,068	3,882E-10
Filip1l	38,234	0,093	804,490	15,377	0,048	-4,395	1,519E-08
PPase †	6,548	0,038	142,819	5,251	0,046	-4,447	1,608E-05
Gas5	0,273	0,018	6,403	16,405	0,043	-4,551	5,693E-14
Pdcd5	0,259	0,013	7,187	6,993	0,036	-4,793	8,780E-13
Btg3	1,125	0,001	190,894	12,644	0,006	-7,407	5,575E-06
9930012K11Rik	1,213	0,000	443,733	15,944	0,003	-8,515	1,324E-15
Cdkn1c	0,518	0,000	247,039	10,957	0,002	-8,896	2,039E-12
GTPase †	0,141	0,000	245,822	15,659	0,001	-10,772	3,212E-07

† TPRp, PPase and GTPase are anonymous replacements for characterised proteins due to possible patent application.
‡ Var: Variance; FC: Fold change

4.2.2 Factor verification using quantitative RT-PCR

In prior to factor verification by quantitative RT-PCR, an appropriate internal control (e.g. a housekeeping gene) had to be chosen for relative comparison. Here, the stable and constant expression at different growth phases is necessary to ensure proper expression differentiation. Therefore, different documented housekeeping genes needed to be measured during a batch culture at different time points (Figure 4.17).

Based on recent publications regarding CHO housekeeping proteins (Bahr et al. 2009), the classical housekeeping gene ActB (actin B) was compared with the adherens junction transmembrane protein Vezt (Vezatin) and Yaf2 (YY1 associated factor 2), which is located within the nucleus. All three proteins are located in different compartments to avoid compartment-specific effects.

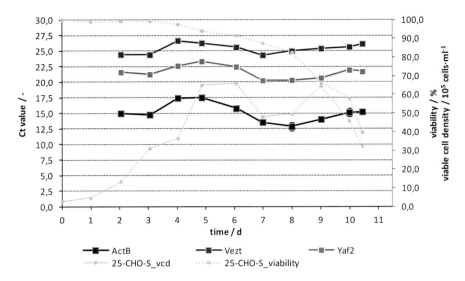

Figure 4.17: Housekeeping gene comparison at different timepoints of 25-CHO-S batch cultivation. The Ct threshold at the normalised fluorescence of 0.100 was determined for Actin B, (ActB), Vezatin (Vezt) and Yaf2 during a 25-CHO-S batch cultivation. Medium: CM1035 + 4 mM Gln. Culture conditions: 37 °C, 80 rpm, 2.5 sL·h^{-1}, 21 % O$_2$, 5 % CO$_2$, 1000 ml spinner flask, 200 ml cell suspension (n = 1). Ct values are mean values of technical duplicates with respective standard deviation. The raw data was kindly provided by Oliver Hermann (former scientist at Celonic GmbH).

As a result, the transcription pattern of ActB (Mean Ct = 15.5), Vezt (Mean Ct = 25.4) and Yaf2 (Mean Ct = 21.6) showed no significant difference, even at very low viabilities. All factors were verified as housekeeping genes. Therefore, ActB was designated as internal standard due to its low Ct values at the quantitative RT-PCR analysis and a slight non-significant chance in transcription between CHO-S-E400/1.104 and 25-CHO-S/2.5C8 seen in HiSeq 2000 NGS transcriptomic analysis ($\log_2$(FC) = 1.238; p-value = 7.66·10^{-3}).

Subsequently, each selected transcript from NGS transcriptomic analysis (Table 4.1) was analysed with the respective primer pair (Table 3.8) by quantitative RT-PCR (Figure 4.18). Here, identical cell pellet aliquots were applied for verification as used for external NGS transcriptomic analysis (Figure 4.16, day 12; Feed: IS CHO Feed-CD XP). In addition, triplicates from CHO CD EfficientFeed A/B fedbatches were applied as well, to avoid or detect feed medium specific effects.

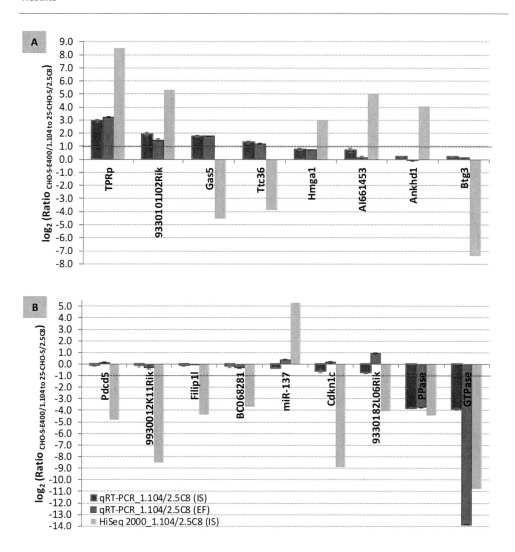

Figure 4.18: Quantitative RT-PCR results compared with HiSeq 2000 NGS transcriptome analysis. Significantly deregulated factors (-1.00 > $\log_2$(Ratio$_{1.104/2.5C8}$) > 1.00) in both feed media were the upregulated factors TPRp, 9330101J02Rik (LOC100759461), Gas5 and Ttc36 (A) as well as the anonymised downregulated factors PPase and GTPase (B). Each qRT-PCR bar represents a normalised mean of technical duplicates as well as biological triplicates against ActB with the resulting standard deviation (error progression) (n = 3·2). The transcriptomic NGS data represent the visualisation of raw data in table 4.1. IS/EF: Cell pellet samples from IS CHO Feed-CD XP or CHO CD EfficientFeed A/B fedbatch, respectively (at day 12; Figure 4.16).

Applying the threshold mentioned above, the oncogenic protein TPRp, the functionally unknown protein 9330101J02Rik (herein called LOC100759461 for *C. griseus* homologue), the non-coding RNA (ncRNA) Gas5 as well as Ttc36 (HBP21) were significantly upregulated in

CHO-S-E400/1.104 at late exponential phase and independent on feed medium applied (log_2(Ratio$_{1.104/2.5C8}$) of TPRp/LOC100759461/Gas5/Ttc36: 2.95 ± 0.10/1.92 ± 0.14/1.76 ± 0.07/1.30 ± 0.13 in IS-fed cultures and 3.21 ± 0.08/1.46 ± 0.13/1.77 ± 0.02/1.18 ± 0.04 in EF-fed cultures).

On the other hand, the membrane-associated protein phosphatase PPase (anonymised) and the src- and fak-associated GTPase (anonymised) were massively downregulated in both feed media (log_2(Ratio$_{1.104/2.5C8}$) of PPase/GTPase: -3.81 ± 0.09/-3.88 ± 0.14 in IS-fed cultures and -3.75 ± 0.03/-13.86 ± 0.09 in EF-fed cultures).

Interestingly, another functionally unknown protein called 9330182L06Rik possessed a strongly feed medium dependent expression. Here, in IS-fed cultures 9330182L06Rik was insignificantly downregulated (log_2(Ratio$_{1.104/2.5C8}$) = -0.73 ± 0.17) and insignificantly upregulated (log_2(Ratio$_{1.104/2.5C8}$) = 0.94 ± 0.06) in EF-fed cultures. These findings reflect the necessity of feed medium comparison.

The insignificantly deregulated factors showed no (Ankhd1, Btg3, Pcdc5, 9930012K11Rik, Filip1l, miR-137, Cdkn1c, 9330182L06Rik) or very poor correlation (Hmga1, Al661453, BC068281) between qRT-PCR and transcriptomic results. The significantly deregulated factors TPRp, 9330101J02Rik/LOC100759461, PPase and GTPase showed a better qRT-PCR/transcriptome correlation, whereat the results for the factors Gas5 and Ttc36 were reciprocal correlated (Figure 4.18). Overall, the quantitative RT-PCR results are poorly comparable with the results of NGS transcriptome analysis, but more reliable (Chapter 5).

Although the differential expression of the factors at the respective maximal cell densities was the most important point of view and crucial for final factor selection, analysis of the samples at exponential phase were interesting as well to elucidate the course of factor transcription. Therefore, the exponential phase samples of both cell lines and feed media (Figure 4.16, day 4) were compared with the already attained results (Figure 4.19).

Here, the ncRNA Gas5 (Growth arrest specific 5) was identically transcribed in exponential as well as early stationary phase, what repudiated the assumption of involvement of Gas5 in growth arrest, since CHO-S-E400/1.104 possessed a improved growth characteristic over 25-CHO-S/2.5C8 (Figure 4.16).

The proteins TPRp, PPase as well as GTPase showed same directed deregulation in exponential and early stationary phase. During exponential phase, TPRp and PPase showed a lower extent in deregulation in mutant cell line CHO-S-E400/1.104 (Figure 4.19).

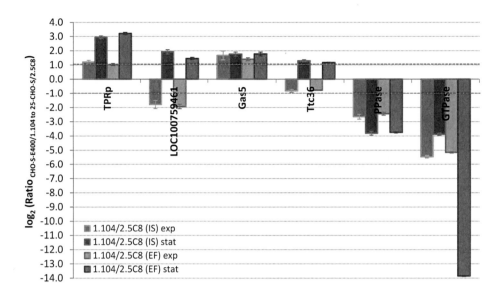

Figure 4.19: Differential expression of putative high cell density factors and comparison between exponential (exp) and early stationary phase (stat). TPRp, Gas5, PPase and GTPase were significantly deregulated in both feed media as well as growth phases (exp and stat) at the same direction with similar or lower extent. The high cell density factors LOC100759461 (9330101J02Rik) and Ttc36 were downregulated at exponential phase and upregulated at early stationary phase. Each bar represents a normalised mean (against ActB) of technical qRT-PCR duplicates as well as biological triplicates with the resulting standard deviation (error progression) (n = 3·2). IS/EF: Cell pellet samples from IS CHO Feed-CD XP or CHO CD EfficientFeed A/B fedbatch, respectively (at day 4 and 12, Figure 4.16). Significance at -1.00 > $log_2(Ratio_{1.104/2.5C8})$ > 1.00.

Interestingly, both tetratricopeptide-containing proteins Ttc36 and LOC100759461 were downregulated in CHO-S-E400/1.104 cells during exponential growth phase (significance for LOC100759461) and significantly upregulated during early stationary phase (Figures 4.18 and 4.19). This similarity underlines a putative tetratricopeptide-containing protein-specific mechanism. Nevertheless, since both proteins were significantly upregulated in early stationary phase, it was assumed that the increased presence of Ttc36 and LOC100759461 during this later growth phase is crucial for resistance at high cell density.

Overall, all factors were robust in expression in terms of absence of any feed dependent effects and therefore were directly connected to stable mutagenesis dependent modification (Chapter 4.1.9). For validation (Chapter 4.3), TPRp, LOC100759461, Gas5 as well as Ttc36 should be constitutively overexpressed in control cell line 25-CHO-S/2.5C8 for confirmation of high cell density inhibition uncoupling effect and in a hIgG-producing cell line to elucidate the effect on its productivity.

The two anonymised, in CHO-S-E400/1.104 downregulated proteins PPase and GTPase should be knocked down or knocked out. Nevertheless, due to further investigations and to not disturb patent application processes, the results of these proteins are not further presented. Further detailed discussion on the putative function of both proteins is performed in chapter 5.

4.3   Factor validation in non-producing and producer cell lines

The validation of identified and verified high cell density (HCD) factors (TPRp, LOC100759461, Gas5 and Ttc36) was performed by constitutive, enhanced CMV driven overexpression in 25-CHO-S/2.5C8 (control clonal progenitor cell line, chapter 4.1.6) and in the clonal high producer CHO cell line K20-3 producing the humanised monoclonal antibody IgG$_B$ (hIgG$_B$, hereinafter called hIgG). The latter cell line possesses resistance against G418/Geneticin (Chapter 3.1), was single cell cloned and selected for maximal hIgG titer and productivity. Therefore, HCD factor cDNA-carrying expression vectors (CV121, chapter 3.8.1) were designed for selection against hygromycin B. Transcription and expression of the protein factors TPRp, LOC100759461 and Ttc36 were directly connected with hygromycin B resistance gene via an internal ribosomal entry site (IRES) to ensure coupled and dependent expression. For the ncRNA Gas5 and its derivates a separate cassette (Table 3.21) was inserted and was not directly controlled by hygromycin B resistance.

Through the subsequent validation by overexpression, the putative maximal cell density as well as growth rate increasing effect of the verified factors should be elucidated. In addition, by overexpression in CHO-K1 derived producer cell line K20-3 the impact on productivity should be observed.

Comparison was performed with HCD factor-overexpressing polyclonal pool cell lines and parallel stably transfected Mock (CV121lin) transfected control polyclonal pool cell line to overcome clonal variation by diverse integration sites and avoid selective pressures during single cell cloning process. In a subsequent step, pool K20-3-derived cell lines with improved characteristics needed to be single cell cloned by limited dilution (Chapter 3.8.4) to further focus on maximal productivity and clonal variation.

For better discrimination from other rodent species, the hamster factor homologue is herein defined with prefix "*cgr*". In addition, viable cell densities and hIgG titers are normalised with the respective control cell line and depicted as perceptual values for better comparison.

4.3.1 Validation of *cgr*TPRp in non-producing 25-CHO-S/2.5C8 cells

The Chinese hamster coding sequence of the anonymised tetratricopeptide-containing protein (*cgr*TPRp) was amplified using cDNA from CHO-S-E400/1.104 and the regarding primer pair (Table 3.14). The sequencing of *cgr*TPRp revealed extent homology of protein sequence to mouse (*mmu*TPRp), rat (*rno*TPRp) and human (*hsa*TPRP) homologues (alignment not shown due to possible patent application). Focussing on percent identity matrix for these species (Figure 4.20), *cgr*TPRp possesses strong coverage and homology to all rodent (*mmu*TPRp and *rno*TPRp) as well as human species (*hsa*TPRP). The murine TPRp (*mmu*TPRp) possesses the strongest phylogenetic relationship to *cgr*TPRp among these species.

		cgr	*mmu*	*rno*	*hsa*
*cgr*TPRp	(experimental sequence)	100.00	92.68	86.99	86.40
*mmu*TPRp	(known NCBI sequence)	92.68	100.00	90.24	86.99
*rno*TPRp	(known NCBI sequence)	86.99	90.24	100.00	82.93
*hsa*TPRp	(known NCBI sequence)	86.40	86.99	82.93	100.00

Figure 4.20: Chinese hamster TPRp (*cgr*TPRp) is highly homologous with TPRp of the other rodent species *M. musculus* (*mmu*TPRp) as well as *R. norvegicus* (*rno*TPRp) and with *H. sapiens* (*hsa*TPRp). The percent identity matrix of TPRp of the alignment reveals the strong relationship between the TPRp protein species.

Stable overexpression of *cgr*TPRp in 25-CHO-S/2.5C8 and polyclonal cell line pool comparison with mock-transfected cell line 25-CHO-S/2.5C8/CV121lin did not show any increase in maximal cell density (Figure 4.21). Furthermore, the polyclonal 25-CHO-S/2.5C8/*cgr*TPRp cell line showed a reduced growth rate during exponential growth phase (day 0 – 4: 87.7 %; day 4 – 5: 94.5 %). Overexpression of *cgr*TPRp was beneficial regarding prolonged culture duration above 80 % viability. Here, shaken culture was significantly prolonged for two days compared to mock-transfected pool cell line.

In addition, cell density did not decrease in stationary phase suggesting the cells possess increased inhibitory mechanisms against apoptosis, autophagy and necrosis. Nevertheless, *cgr*TPRp was not suitable to increase maximal cell density. Therefore, overexpression in K20-3 production cell line was secondary and not performed during this thesis.

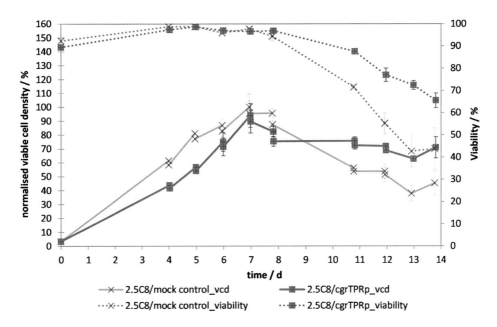

Figure 4.21: Overexpression of *cgr*TPRp in 25-CHO-S/2.5C8 and comparison with Mock cell line (25-CHO-S/2.5C8/CV121lin) in parallel fedbatch cultures. Medium: 87.5 % CM1035 / 10 % CHO CD EfficientFeed A/B + 5 mM Gln + 200 $\mu g \cdot ml^{-1}$ hygromycin B. Feed: 80 % CHO CD EfficientFeed A/B + 40 mM Gln + 200 $\mu g \cdot ml^{-1}$ hygromycin B. Feeding rate: 500 μl at day 4, 5, 6, 7, 11, and 12; 1000 μl at day 8. Culture conditions: 37 °C, 200 rpm, amplitude: 30 mm, 5 % CO_2, aeration: A-D, BFT50, 10 ml cell suspension. Each point represents a mean of biological triplicates with resulting standard deviation (n = 3). Corresponding viable cell density at 100 %: $7.40 \cdot 10^6$ cells$\cdot ml^{-1}$.

4.3.2 Validation of LOC100759461/TPR repeat-containing protein C10orf93-like

4.3.2.1 Structural and homological relationship of C10orf93-like species

The functionally unknown *C. griseus* locus LOC100759461 that encodes for the tetratricopetide repeat-containing protein C10orf93-like (Gene ID: 100759461; GenBank: XP_003510630.1) (herein called *cgr*C10orf93-like) was chosen to be the nearest homologue to *M. musculus* locus 9330101J02Rik (Gene ID: 664857) coding for murine C10orf93 (XP_001472215.2) (hereinafter called *mmu*C10orf93) (Figure 4.22 A), which was originally found in differential NGS transcriptome analysis to be overregulated in the mutant cell line (Chapter 4.2.1). In addition, the functionally unknown tetratricopeptide repeat-containing protein *cgr*C10orf93-like (LOC100759461) was found to be strongly related to C10orf93/Ttc40/TTC40 homologues of other species (Figure 4.22). Furthermore, *cgr*C10orf93-like represents the shortest C10orf93/Ttc40 homologue in this list. The putative function and cellular localisation is discussed in chapter 5.3.3.

A

		cgr	mmu	rno	hsa
cgrC10orf93-like	(XM_003510582.1)	100.00	85.58	85.97	78.08
mmuC10orf93	(XM_001472165.2)	85.58	100.00	90.33	74.50
rnoTtc40-like	(XM_002725726.2)	85.97	90.33	100.00	72.90
hsaTTC40	(NM_001200049.2)	78.08	74.50	72.90	100.00

B

		cgr	mmu	rno	hsa
cgrC10orf93-like	(XP_003510630.1)	100.00	79.89	80.52	72.55
mmuC10orf93	(XP_001472215.2)	79.89	100.00	85.12	67.29
rnoTtc40-like	(XP_002725772.2)	80.52	85.12	100.00	62.31
hsaTTC40	(NP_001186978.2)	72.55	67.29	62.31	100.00

Figure 4.22: Percent identity matrix of coding sequence (CDS) (A) and protein sequence (B) of TPR repeat-containing protein homologues of cgrC10orf93-like. All homologues have transmembrane domaines with strong variations at intracellular C-terminus. The C. griseus C10orf93 homologue cgrC10orf93-like possesses strong homology with mmuC10orf93 (9330101J02Rik) rnoTtc40-like and hsaTTC40 in CDS as well as protein sequence. Percent identity matrices were calculated using ClustalW2/ClustalΩ using the gene identification codes shown.

Since amplification of cgrC10orf93-like by preparative PCR (Chapter 3.7.2) was not successful, a codon-optimised variant was synthesised, cloned (Table 3.21) and overexpressed in 25-CHO-S/2.5C8 as well as compared with the mock-transfected cell line (Figure 4.23).

4.3.2.2   Validation of cgrC10orf93-like in 25-CHO-S/2.5C8 and K20-3 cell line

Overexpression of cgrC10orf93-like/LOC100759461 in 25-CHO-S/2.5C8 led to abolished cell growth and cell lysis at later culture phase. Beside these results, cgrC10orf93-like/LOC100759461 was transfected in parallel to the transfection in 25-CHO-S/2.5C8 and overexpressed in mAb-producing K20-3 cell line to observe a possible beneficial effect on productivity.

Surprisingly, overexpression of cgrC10orf93-like/LOC100759461 in K20-3 cell line did inhibit growth rate in batch culture in a lower extent than in 25-CHO-S/2.5C8 cell line (Figure 4.24). The maximal cell density reached was 79.1 ± 2.3 % of the maximal cell density of control mock-transfected cell line (K20-3/CV121lin) ($4.30 \cdot 10^6 \pm 0.13 \cdot 10^6$ cells·ml^{-1}). In addition, the mAb-titer at a viability ≥ 75 % was insignificantly increased (1.7 %), whereas cell specific productivity (csp) decreased to 89.1 % compared to the mock control csp. Nevertheless, the duration of culture with viable cells above a viability of 78 % was significantly prolonged for two days. Therefore, overexpression of cgrC10orf93-like/LOC100759461 prolonged viable cell culture at least for the K20-3 cell line.

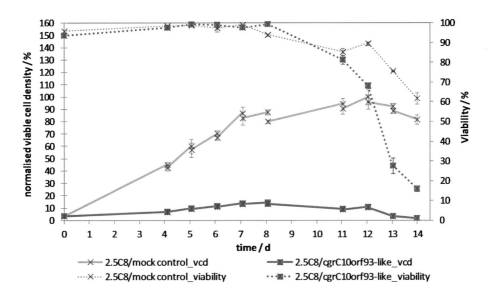

Figure 4.23: Overexpression of *cgr*C10orf93-like/LOC100759461 in 25-CHO-S/2.5C8 and comparison with Mock cell line (25-CHO-S/2.5C8/CV121lin) in parallel fedbatch cultures. Medium: 87.5 % CM1035 / 10 % CHO CD EfficientFeed A/B + 5 mM Gln + 200 µg·ml^{-1} hygromycin B. Feed: 80 % CHO CD EfficientFeed A/B + 40 mM Gln + 200 µg·ml^{-1} hygromycin B. Feeding rate: 500 µl at day 4, 5, 6, 7, 11, 12 and 13; 1000 µl at day 8. Culture conditions: 37 °C, 200 rpm, amplitude: 30 mm, 5 % CO_2, aèration: A-D, BFT50, 10 ml cell suspension. Each point represents a mean of biological triplicates with resulting standard deviation (n = 3). Corresponding viable cell density at 100 %: $7.08·10^6$ cells·ml^{-1}.

Another result of *cgr*C10orf93-like/LOC100759461 overexpression was the nearly abolished aggregation of K20-3/*cgr*C10orf93-like (Figure 4.25). The K20-3 cells as well as the polyclonal mock-transfected K20-3 cell line tended to strong aggregation. Here, K20-3/mock cell line showed a grade of aggregation of approximately 80 %. Depending on growth phase, culture duration and culture vessel the grade of aggregation could vary between approx. 10 % and 80 % (data not shown). In contrast to mock cell line, K20-3/*cgr*C10orf93-like showed a grade of aggregation of approximately 10 % (Figure 4.25) and maximal 20 %, respectively.

In summary, overexpression of *cgr*C10orf93-like/LOC100759461 in 25-CHO-S/2.5C8 was not beneficial at all. Here, the reduction of aggregation could not be observed since this cell line was selected for reduced aggregation (Chapter 4.1.6). On the other hand, *cgr*C10orf93-like/LOC100759461 was suitable for prolong the viable culture duration as well as reduced the cell aggregation once transfected and overexpressed in K20-3 without greatly affecting the mAb production.

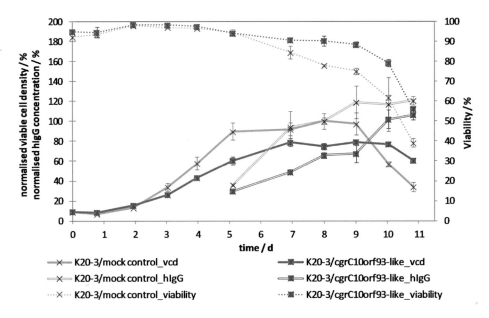

Figure 4.24: Overexpression of *cgr*C10orf93-like/LOC100759461 in mAb-producing K20-3 cell line and comparison with Mock cell line (K20-3/CV121lin) in parallel batch cultures. Medium: 87.5 % CM1035 / 10 % CHO CD EfficientFeed A/B + 5 mM Gln + 200 µg·ml^{-1} hygromycin B. Culture conditions: 37 °C, 200 rpm, amplitude: 30 mm, 5 % CO_2, aeration: A-D, BFT50, 10 ml cell suspension. Each point represents a mean of biological triplicates with resulting standard deviation (n = 3). Corresponding viable cell density at 100 %: 5.43·10^6 cells·ml^{-1}. Corresponding hIgG concentration (titer) at 100 %: 146.8 mg·l^{-1} at a viability of 78.0 ± 0.5 %.

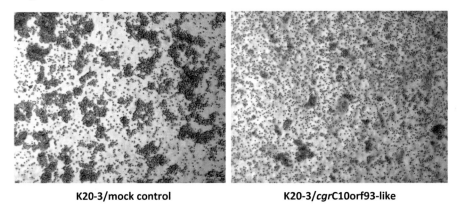

Figure 4.25: Overexpression of *cgr*C10orf93-like/LOC100759461 in mAb-producing K20-3 cell line led to reduced cell aggregation over mock control. Medium: 87.5 % CM1035 / 10 % CHO CD EfficientFeed A/B + 5 mM Gln + 200 µg·ml^{-1} hygromycin B. Culture conditions: 37 °C, 5 % CO_2, 6-well plate, 2 ml cell suspension. The pictures are representative for the respective cell line and were generated six days after seeding.

4.3.3 Validation of *cgr*Gas5 variants and *cgr*Snord78

In following, the isolation and validation of *C. griseus* full-length Gas5 (growth arrest specific 5), its splice variants as well as intronic sequences (*cgr*Snord78) are described in detail.

4.3.3.1 Identification, isolation and characterisation of *cgr*Gas5 variants

The constant upregulation of Gas5 in CHO-S-E400/1.104 (Figure 4.18) was surprising since this non-coding RNA (ncRNA) species was recently documented to be growth arrest associated (Kino *et al.* 2010; Mourtada-Maarabouni *et al.* 2010).

For elucidation of the *C. griseus* sequence of Gas5 (*cgr*Gas5), the qRT-PCR amplicon (using *mmu/rno* homologous Gas5-primer in table 3.8) was sequenced (Chapter 3.9.3) to subsequently blast the sequence against the CHO genome database (Hammond *et al.* 2012a). Here, the results revealed very strong coverage (100% under disregard of primer sequences) to the whole genome shotgun sequences (wgs) *Cricetulus griseus* scaffold172_47 (GenBank: AFTD01011934.1) as well as C41583244_1 (GenBank: AFTD01262011.1; gi|342371527) (Score: 223; Identities (query length): 123/123 (125); Expect: 2e-56) (Figure 4.26).

The latter wgs sequence, which was not declared as Gas5 yet, possesses 2029 basepairs (bp) and therefore is shorter than other rodent Gas5 and human GAS5 species (Table 4.2).

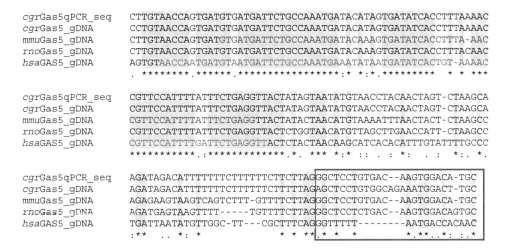

Figure 4.26: DNA sequence alignment of qRT-PCR amplicon (*cgr*Gas5qPCR_seq) with four genomic Gas5/GAS5 species (Table 4.2). Since the primer set was designed by homology comparison between *mmu*Gas5 and *rno*Gas5 genomic region, the amplicon did not possess 100 % coverage to *cgr*Gas5 genomic DNA. Highlighted are the consensus sequences (grey rear), the Gas5/GAS5 exon sequence (framed in red) as well as the documented Snord47/SNORD47 sequences (red letters).

Table 4.2: Gas5/GAS5 species and their documented components

	*hsa*GAS5	*mmu*Gas5	*rno*Gas5	*cgr*Gas5
Length of genomic sequence	4087 bp (gi|29498337) (Gene ID: 60674)	3327 bp (gi|34740397) (Gene ID: 14455)	3322 bp (gi|389669371) (Gene ID: 81714)	2029 bp (gi|342371527) (experimental)
Length of spliced ncRNA	651 bp (gi|144226237)	2554 bp (gi|144226204)	429 bp (gi|84871984)	-
Number of documented intronic sequences	10	3	10	-
Documented intronic C/D Box snoRNA species In order 5' → 3'	SNORD74	-	-	-
	SNORD75	-	-	-
	SNORD76	-	-	-
	SNORD77	-	-	-
	SNORD44	-	-	-
	SNORD78	-	-	-
	SNORD79	-	-	-
	SNORD80	-	-	-
	SNORD47	Snord47	-	-
	SNORD81	-	-	-
Documented intronic microRNA spezies	-	*mmu*-miR-5117 dead entry (Castellano and Stebbing 2013)	-	-

Overall, the coverage and homology of full-length Gas5/Gas5 sequences are poor (Figure 4.27). Interestingly, intronic small nucleolar RNA C/D Box (Snord/SNORD) sequences were more conserved than the exon regions (Figure 4.26, representatively for all analysed sequences possessing Snords/SNORDs). In addition, the alignment of full-length genomic Gas5/GAS5 DNA (data not shown) showed that the rodent Gas5 species were truncated at their 5'-ends resulting in shorter RNA/DNA sequences compared to *hsa*GAS5.

		cgr	*mmu*	*rno*	*hsa*
*cgr*Gas5	(gi|342371527)	100.00	72.02	70.05	60.62
*mmu*Gas5	(gi|34740397)	72.02	100.00	80.09	59.50
*rno*Gas5	(gi|389669371)	70.05	80.09	100.00	58.84
*hsa*GAS5	(gi|29498337)	60.62	59.50	58.84	100.00

Figure 4.27: Percent identity matrix of genomic Gas5/GAS5 sequences. All Gas5/GAS5 species showed distinct differences in their sequences. Percent identity matrices were calculated using ClustalW2/ClustalΩ and the gene identification codes shown.

Based on this alignment study, the wgs sequence C41583244_1 (GenBank: AFTD01262011.1, gi\|342371527) was designated as the genomic sequence for *cgr*Gas5 and therefore this sequence was amplified by PCR using the primers flanking this sequence (Table 3.14). To elucidate the impact of the Snord containing intronic sequences, the *cgr*Gas5 full-length variant and splice variants were generated by using the genomic DNA of CHO-S-E400/1.104 (Chapter 3.7.1) as well as the cDNA from qRT-PCR verification (Chapter 3.6.2, using the poly dT primer) as templates, respectively. Here, the full-length *cgr*Gas5 with 100 % coverage to wgs sequence C41583244_1 was isolated as well as two different splice variants (Table 4.3).

Table 4.3: Comparison between full-length *cgr*Gas5 and its isolated splice variants.

	*cgr*Gas5_2029bp	*cgr*Gas5_1237bp	*cgr*Gas5_788bp
Length genomic sequence	2029 bp	1237 bp	788 bp
Template for isolation	gDNA (CHO-S-E400/1.104)	cDNA (CHO-S-E400/1.104)	cDNA (CHO-S-E400/1.104)
Number of intronic sequences: observed / theoretical	4 / 5 [†]	1 / 2 [†]	0 / 1 [†]
C/D Box snoRNA species In order 5' → 3'	Snord44 Snord78 Snord79 Snord80[‡] Snord47	Snord44 Snord78 - - -	Snord44 - - - -

[†] Snord44 (*cgr*Snord44) is flanked by intronic consensus sequences but splicing was not observed.

[‡] The hamster intronic sequence possessing *cgr*Snord80 displays strong homology to the human SNORD77 and SNORD80 sequences (Table 4.4).

The full-length *cgr*Gas5 (*cgr*Gas5_2029 bp) possessed five C/D Box small nucleolar RNA (Snord) species (*cgr*Snord44, *cgr*Snord78, *cgr*Snord79, *cgr*Snord77/80 or snR60/Z15/Z230/Z193/J17 as well as *cgr*Snord47). The two isolated splice variants possessed the C/D Box snoRNA species *cgr*Snord44 as well as *cgr*Snord78 or *cgr*Snord44, respectively (Table 4.3). The alignment of *cgr*Gas5 C/D Box snoRNA with their documented homologues revealed a strong homology of the putative C/D and C'/D' Box consensus sequences (Table 4.4, van Nues *et al.* 2011).

Four differences of hamster *cgr*Gas5 to human and rodent species could be shown: (1) The full-length species is massively truncated compared to other Gas5/GAS5 homologues; (2) In *cgr*Gas5 the number of putative functional C/D box snoRNAs (six intronic snoRNAs, Table 4.3) is reduced compared to i.e. *hsa*GAS5 (Gene ID: 60674) with ten intronic snoRNAs by truncation at 5'-terminal (absence of hamster Snord74, Snord75 and Snord76) as well as 3'-terminal sequences (absence of hamster homologue for Snord81); (3) The present data shows a theoretical splicing ability for *cgr*Snord44, but could not be observed. Furthermore, it was suggested that the *cgr*Snord78 containing intron in splice variant *cgr*Gas5_1237bp has a poor ability to be spliced. Therefore, the splice variant *cgr*Gas5_1237bp was able to be isolated in addition to the putatively fully spliced *cgr*Gas5_788bp splice variant.

All presented elements are roughly documented under gi|344151080, whose entry is identical to gi|342371527, but with further sequence information. For the first time, the splicing ability of snoRNA containing *cgr*Gas5 introns are presented. The hitherto documented small nucleolar RNA snR60/Z15/Z230/Z193/J17 (gi|344151080) is stated as *cgr*Snord80 showing strong homology to *hsa*SNORD80 (Table 4.4).

Table 4.4: DNA sequence alignment of *cgr*Gas5 snoRNAs with documented homologues. All hamster sequences derived from NCBI entry gi|344151080. Putative C/D Box consensuses (Smith and Steitz 1998) are shown in red (light red: C sequences; dark red: D'/D sequences). The C/D box snoRNA species are shown in order of their appearance 5′ → 3′.

```
cgrSnord44                 CCUGCAUGAUGACAAGCAAAUGCUGACU-AACAU
hsaSNORD44  (NR_002750.2)  CCUGGAUGAUGAUAAGCAAAUGCUGACUGAACAU
                           **** ******* ************** *****

cgrSnord44                 GAAGCUCUUAAUUAGCUCUAUCUGAUC
hsaSNORD44  (NR_002750.2)  GAAGGUCUUAAUUAGCUCUAACUGACU
                           **** *****************:****


cgrSnord78                 UUGUAAUGAUGUUGAUCCAAAUGUCUGACCUGAAAAU
hsaSNORD78  (NR_003944.1)  GUGUAAUGAUGUUGAU-CAAAUGUCUGACCUGAAAUG
                           **************** ******************:

cgrSnord78                 AACAUACAUGUAGACAGCAAAUUAAACACUGAAGAA
hsaSNORD78  (NR_003944.1)  AG----CAUGUAGACA--AAGGUA------------
                           *.     *********   **. **


cgrSnord79                 UACUGUUAGUGAUGAUGUAUAAAGCUAAACAGAUGGGAAUCUC
hsaSNORD79  (NR_003939.1)  --CUGUUAGUGAUGAUUUUUAAAAUUAAAGCAGAUGGGAAUCCC
                           ************** *:.*** :**.************ *

cgrSnord79                 UCUGAAUAAG--AUUGAAGAUUAAUUCUUAAGCUGAAACAGUA
hsaSNORD79  (NR_003939.1)  UCUGAGAAAGAAAUGGAGAUUAAU-CUUAAACUGAAACAG--
                           *****.:*** *:**.********* ***** *********


cgrSnord80                 CCGAUACUGUGAUGAUAACAUAGUUCAGCAGACUUAACCUG
hsaSNORD77  (NR_003943.1)  CAGAUACUAUGAUGGUUGCAUAGUUCAGCAGAUUUAAUC--
                           *.******.*****.*: *************** **** *

cgrSnord80                 AUGAACAAUCCUAAGUCUUUCGCUCCUAUCUGACGUAUCUG
hsaSNORD77  (NR_003943.1)  AUGAAGAG----AUG--------UACUAUCUGUC-------
                           ***** *.      *.*       * .*******:.*


cgrSnord80                 CCGAUACUGUGAUGAUAACAUAGUUCAGCAGACUUAACCUG
hsaSNORD80  (NR_003940.1)  -----ACAAUGAUGAUAACAUAGUUCAGCAGACUAACGCUG
                           **:.************************:*.  ***

cgrSnord80                 AUGAACAAUCCUAAGUCUUUCGCUCCUAUCUGACGUAUCUG
hsaSNORD80  (NR_003940.1)  AUGAGCAAUAUUAAGUCUUUCGCUCCUAUCUGAUG------
                           ****.****. *******************. *


cgrSnord47                 AACCAGUGAUGUGAUGAUUCUGCCAAAUGAUACAUAGUGA
hsaSNORD47  (NR_002746.1)  AACCAAUGAUGUAAUGAUUCUGCCAAAUGAAAUAUAAUGA
mmuSnord47  (NR_028543.1)  ----------GUGAUGAUUCUGCCAAAUGAUACAAAGUGA
                           ****************** * * ***

cgrSnord47                 UAUCACCUUUAAAACCGUUCCAUUUUAUUUCUGAGGUUA
hsaSNORD47  (NR_002746.1)  UAUCAC-UGUAAAACCGUUCCAUUUUGAUUCUGAGGUU-
mmuSnord47  (NR_028543.1)  UAUCACCUUUAAA-CCGUUCCAUUUUAUUUCUGAGG---
                           ****** * **** ************  ********
```

Interestingly, the sequences between both C/C' box elements in *cgr*Snord77 and *cgr*Snord80 are identical and predicted to mediate the 2'-*O*-methylation of 28S ribosomal subunit (Smith and Steitz 1998). Nevertheless, this intronic snoRNA is further mentioned as *cgr*Snord80.

The next step involves cloning as well as validation of all three *cgr*Gas5 species in progenitor cell line 25-CHO-S/2.5C8 to observe their behaviour during constitutive overexpression in non-producing CHO cell lines and putative functions of their intronic or ncRNA sequences, respectively. Further information about the differences in the Gas5/GAS5 species as well as the contents in *cgr*Gas5 are presented in chapter 5.4.1 and figure 5.15.

4.3.3.2   Validation of *cgr*Gas5 variants in 25-CHO-S/2.5C8

The preparative PCR fragments (Chapter 3.7.2 with primers in table 3.14) of the *cgr*Gas5 variants   (*cgr*Gas5_2029bp/*cgr*Gas5,   *cgr*Gas5_1237bp/*cgr*Gas5$_{Snord\Delta79\Delta77/80\Delta47}$ and *cgr*Gas5_788bp/*cgr*Gas5$_{Snord\Delta78\Delta79\Delta77/80\Delta47}$) were subsequently subcloned within the CV001 basal vector excising the IntronA-MCS-pA$_{BGH}$-ori$_{F1}$-P$_{SV40}$-G418^R with *Bsm*BI and *Psi*I (Chapter 3.8.1). The resulting *cgr*Gas5 variants flanking CMV (upstream) and pA$_{SV40}$ (downstream) were subsequently isolated and cloned within the semi-blunt CV121 vector upstream of the P$_{CMV}$-IntronA-MCS-IRES$_{FMDV}$-EM7-Hygro^R-pA$_{SV40}$ expression cassette (Table 3.21). The resulting three expression vectors (CV142, CV143 and CV144, respectively) were linearised with *Bsp*HI to excise the Ampicillin resistance (Amp^R) (Chapter 3.7.12) and allow defined genomic integration events.

Subsequently, the linearised vector constructs as well as the in parallel linearised mock control CV121lin were introduced into the monoclonal progenitor cell line 25-CHO-S/2.5C8. Here, due to the separated expression cassettes, the finally transcribed ncRNA of the *cgr*Gas5 variants were not directly controlled by hygromycin B resistance.

The transfected 25-CHO-S/2.5C8 polyclonal cell line pools were compared to observe the change in cell density and overall growth as well as viability behaviour. Overall, constitutive overexpression of the *cgr*Gas5 variants in 25-CHO-S/2.5C8 (2.5C8/*cgr*Gas5_2029bp, 2.5C8/*cgr*Gas5_1237bp and 2.5C8/*cgr*Gas5_788bp) did not significantly increase the maximal cell density but was able to improve the viability characteristics compared to mock-transfected control 25-CHO-S/2.5C8/CV121lin (2.5C8/mock control) (Figure 4.28 A). Moreover, the overexpression of either the full length *cgr*Gas5 variant *cgr*Gas5_2029bp or the putatively fully spliced variant *cgr*Gas5_788bp (Chapter 4.3.3.1) reduced the maximal cell density when compared with the mock control cell line. Nevertheless, this effect was at least abolished regarding the integral viable cell density above a viability of 70 % (Table 4.5). This effect could be observed due to the prolonged cultivation duration in 2.5C8/*cgr*Gas5_2029bp as well as 2.5C8/*cgr*Gas5_788bp, i.e. the viable culture over viabilities of 70 % was prolonged for two days.

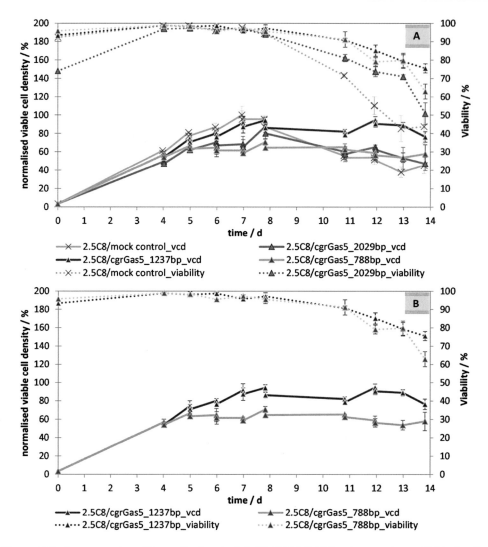

Figure 4.28: Overexpression of the *cgr*Gas5 variants in 25-CHO-S/2.5C8 and comparison with Mock cell line (25-CHO-S/2.5C8/CV121lin) in parallel fedbatch cultures (A). Distinct comparison of *cgr*Gas5_1237bp/*cgr*Gas5$_{Snord\Delta79\Delta77/80\Delta47}$ and *cgr*Gas5_788bp/*cgr*Gas5$_{Snord\Delta78\Delta79\Delta77/80\Delta47}$ (B). Medium: 87.5 % CM1035 / 10 % CHO CD EfficientFeed A/B + 5 mM Gln + 200 µg·ml^{-1} hygromycin B. Feed: 80 % CHO CD EfficientFeed A/B + 40 mM Gln + 200 µg·ml^{-1} hygromycin B. Feeding rate: 500 µl at day 4, 5, 6, 7, 11, 12 and 13; 1000 µl at day 8. Culture conditions: 37 °C, 200 rpm, amplitude: 30 mm, 5 % CO_2, aeration: A-D, BFT50, 10 ml cell suspension. Each point represents a mean of biological triplicates with resulting standard deviation (n = 3). Corresponding viable cell density at 100 %: 7.40·10^6 cells·ml^{-1} (Maximal viable cell density of mock control).

Table 4.5: Changes in integral viable cell density (ivcd) of the three isolated *cgr*Gas5 variants overexpressed 25-CHO-S/2.5C8 compared to the mock control cell line. The data points (only mean values) with viabilities above 70 % were regarded (corresponding Figure 4.28).

Stably transfected cell line population compared to mock control cell line	ivcd [%]	viability [%]	day [d]
2.5C8/mock control	100.0	71.6	11
2.5C8/*cgr*Gas5 (2.5C8/*cgr*Gas5_2029bp)	114.9	70.9	13
2.5C8/*cgr*Gas5$_{Snord\Delta79\Delta77/80\Delta47}$ (2.5C8/*cgr*Gas5_1237bp)	155.3	75.5	14
2.5C8/*cgr*Gas5$_{Snord\Delta78\Delta79\Delta77/80\Delta47}$ (2.5C8/*cgr*Gas5_788bp)	111.7	79.8	13

Interestingly, the overexpression of the partially spliced *cgr*Gas5 variant *cgr*Gas5_1237bp in 25-CHO-S/2.5C8 prolonged the viable cell culture (≥70 % viability) for three days and significantly increased the integral viable cell density (+55 %) when compared with the mock cell line (Table 4.5). The maximal cell density was not changed significantly, but after eight days in culture the 25-CHO-S/2.5C8/*cgr*Gas5_1237bp cell line pool maintained the viable cell density and high viabilities (Figure 4.28 A). Therefore, this *cgr*Gas5 variant was suitable for slightly improve a CHO cell line once overexpressed.

Regarding the splice variant *cgr*Gas5_1237bp as well as the putatively fully spliced *cgr*Gas5 variant *cgr*Gas5_788bp, whose differ in Snord78 (*cgr*Snord78) containing intron sequence only (Table 4.3), the characteristics for viability and viable cell density maintenance were comparable (Figure 4.28 B). In addition, comparison of both splice variants showed a significant difference in maximal cell density (up to 65 % difference, day 13, >79 % viability) as well as in integral viable cell density (30 % difference, day 13, >79 % viability) (Figure 4.28 B).

Both *cgr*Gas5 splice variants, *cgr*Gas5_1237bp and *cgr*Gas5_788bp, differed in the occurrence of a single intron possessing the small nucleolar C/D box RNA *cgr*Snord78. This fact led to the assumption that the presence of the intronic *cgr*Snord78 was suitable to shift and increase the maximal cell density (Figure 4.28 B). Hence, overexpression of *cgr*Snord78 could increase the maximal cell density without affecting the viability characteristics significantly.

Therefore, the *cgr*Snord78-containing *cgr*Gas5-intron should further be isolated and characterised regarding the effect on cell density and productivity (Chapter 4.3.3.3).

4.3.3.3   Isolation and validation of *cgr*Snord78 in 25-CHO-S/2.5C8 as well as K20-3 cell line

Using the vector for *cgr*Gas5$_{Snord\Delta79\Delta77/80\Delta47}$ (*cgr*Gas5_1237bp) expression (CV143) and the regarding primer pair (Table 3.14, *cgr*Snord78), the full *cgr*Snord78-containing *cgr*Gas5-intron was amplified. Subsequently, the resulting PCR fragment was cloned into CV121 by excising the full IntronA sequence (Table 3.21). The resulting vector (CV155) was further linearised (Chapter 3.7.12) and transfected (Chapter 3.8.2) into 25-CHO-S/2.5C8 in parallel to the linearised mock vector CV121lin. Here, both vectors possessed one intron each with almost identical sizes and different intronic consensus sequences leading to putatively distinct splice efficiencies.

Overexpression of *cgr*Snord78 intron and therefore of the *cgr*Snord78 C/D Box snoRNA in 25-CHO-S/2.5C8 showed no significant effects in growth characteristics (viable cell density, viability and growth rate) in uncontrolled fedbatch culture (Figure 4.29).

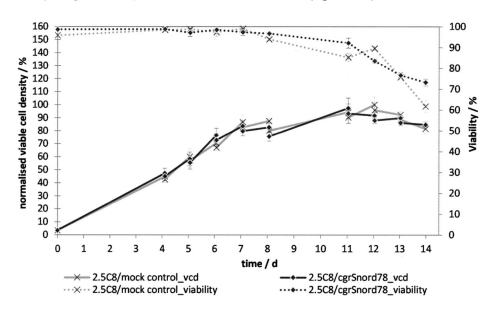

Figure 4.29: Overexpression of intronic *cgr*Snord78 in 25-CHO-S/2.5C8 and comparison with Mock cell line (25-CHO-S/2.5C8/CV121lin) in parallel fedbatch cultures. Medium: 87.5 % CM1035 / 10 % CHO CD EfficientFeed A/B + 5 mM Gln + 200 µg·ml^{-1} hygromycin B. Feed: 80 % CHO CD EfficientFeed A/B + 40 mM Gln + 200 µg·ml^{-1} hygromycin B. Feeding rate: 500 µl at day 4, 5, 6, 7, 11, 12 and 13; 1000 µl at day 8. Culture conditions: 37 °C, 200 rpm, amplitude: 30 mm, 5 % CO_2, aeration: A-D, BFT50, 10 ml cell suspension. Each point represents a mean of biological triplicates with resulting standard deviation (n = 3). Corresponding viable cell density at 100 %: 7.08·10^6 cells·ml^{-1}. (Maximal viable cell density of mock control).

Therefore, this intronic sequence was not toxic to this non-producer cell line. For further understanding of a putative expression enhancing effect of *cgr*Snord78, the linearised expression vector possessing the *cgr*Snord78 intron (CV155lin) was stably overexpressed in the CHO hIgG-producing cell line K20-3. In parallel, the linearised mock vector (CV121lin) was transfected as well.

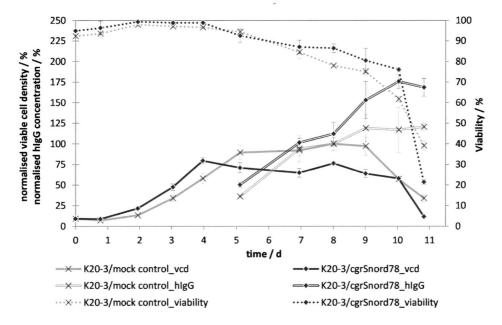

Figure 4.30: Overexpression of *cgr*Snord78-containing intron in mAb-producing K20-3 cell line and comparison with Mock cell line (K20-3/CV121lin) in parallel batch cultures. Medium: CM1035 + 5 mM Gln + 200 µg·ml⁻¹ hygromycin B + 600 µg·ml⁻¹ G418. Culture conditions: 37 °C, 200 rpm, amplitude: 30 mm, 5 % CO_2, aeration: A-D, BFT50, 10 ml cell suspension. Each point represents a mean of biological triplicates with resulting standard deviation (n = 3). Corresponding viable cell density at 100 %: $5.43 \cdot 10^6$ cells·ml⁻¹. Corresponding mean hIgG concentration (titer) at 100 %: 146.8 mg·l⁻¹ at a viability of 78.0 ± 0.5 %.

The overexpression of *cgr*Snord78 in K20-3 cell line slightly increased the growth rate in batch culture but reduced the maximal cell density by earlier stationary phase entry (Figure 4.30). The maximal cell density reached was 79.5 ± 3.9 % ($4.32 \cdot 10^6$ ± $0.21 \cdot 10^6$ cells·ml⁻¹) of the maximal cell density of control mock-transfected cell line (K20-3/CV121lin). In addition, the mAb-titer as well as the cell specific productivity (csp) at a viability ≥75 % were both significantly increased (+ 47.8 % and + 47.5 % respectively). The duration of culture with viable cells above a viability of 75 % was significantly prolonged for one to two days. Therefore, the intronic *cgr*Snord78 prolonged viable cell culture and productivity (titer as well as csp) once stably overexpressed in K20-3 cell line and cultured in batch culture.

To overcome possible nutrition limitation effects, which were suggested due to the comparatively rapid viability decrease in K20-3/*cgr*Snord78 batch culture (Figure 4.30), the culture modus was changed to fedbatch. In addition, two different feed media (CHO CD

EfficientFeed A/B and IS CHO Feed-CD XP) were applied in parallel to observe the efficiencies of both media to putatively increase growth, the maximal cell density as well as the productivity.

Both fedbatch culture processes were carried out using the same feed ratios and schedules (Figure 4.31). Interestingly, regardless the applied feed, the increase in maximal cell density was negotiable and insignificant (< 1.2x). In addition, the viable cell culture of K20-3/mock as well as K20-3/*cgr*Snord78 (≥ 75 % viability) was prolonged about one day. Both fedbatch cultures resulted in similar and insignificant different hIgG titers (436.2 ± 48.9 mg·l^{-1} for the EF-fedbatch and 431.7 ± 94.1 mg·l^{-1} for the IS-fedbatch processes at viabilities over 85 %, respectively) and therefore similar cell specific productivities (11.2 pg·cell^{-1}·d^{-1} and 10.4 pg·cell^{-1}·d^{-1}, respectively; viability ≥80 %). Here, the hIgG titer (viability ≥ 80 %) was increased 2.65-fold (EF) and 2.62-fold (IS) compared to the batch culture, respectively (Figure 4.30). The cell specific productivities (csp) in the respective fedbatches were increased 1.50-fold (EF) and 1.40-fold (IS) (viability ≥80 %), in which the IS-fedbatch reached slightly higher cell densities at almost identical hIgG titers.

The discrepancy in perceptual titer plot shown in figure 4.31 was due to the different response of K20-3/mock control cell line of the feed media. By applying the EF-feed, the titer of the K20-3/mock control cell line was significantly increased compared with the K20-3/mock control IS-fedbatch (1.6-fold increased, viability ≥ 80 %). Therefore, the difference in hIgG titer for K20-3/mock control cell line in EF-feed applied fedbatch was not significantly increased compared to the K20-3/*cgr*Snord78 hIgG titer in fedbatch processes with IS-feed application (Figure 4.31).

Another discrepancy was the rapid degradation of hIgG below viabilities of 80 % (shown as decrease in hIgG titer) in IS-feeded fedbatches. Therefore, for K20-3/*cgr*Snord78 cell lines the application of CHO CD EfficientFeed A/B (EF) was more suitable compared to IS CHO Feed-CD XP (IS). On the other hand, the latter feed (IS) could rather be suitable to select clonal subpopulations (Chapter 4.3.3.4) without degradation capability.

Therefore, it could be shown that the overexpression of intronic *cgr*Snord78 could increase the hIgG production (titer) by increasing the cell specific productivity and by prolonging the viable cell culture. Overexpression of intronic *cgr*Snord78 did neither influence the maximal cell density nor significantly increase the growth rate or reduce the aggregation rate (comparable with or rather higher than K20-3/mock control in figure 4.25, data not shown).

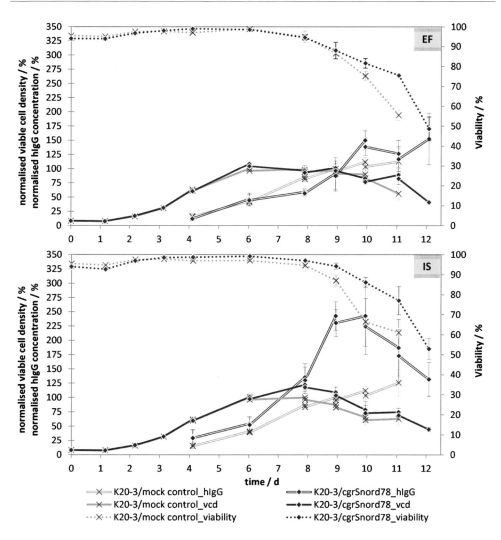

Figure 4.31: Overexpression of *cgr*Snord78-containing intron in mAb-producing K20-3 cell line and comparison with Mock cell line (K20-3/CV121lin) in parallel fedbatch cultures using CHO CD EfficientFeed A/B (EF) or IS CHO Feed-CD XP (IS). Medium: CM1035 + 5 mM Gln + 200 µg·ml^{-1} hygromycin B + 600 µg·ml^{-1} G418. Applied feed (EF or IS) composition: 80 % Feed + 40 mM Gln + 200 µg·ml^{-1} hygromycin B + 600 µg·ml^{-1} G418. Feeding rate (corresponding to L-Glutamine end concentration): 1.5 mM at day 4, 6 and 8; 2.0 mM at day 9; 3.0 mM at days 10 and 11. Total feed (EF or IS) intake (cumulative): 22.0 % at day 11. Culture conditions: 37 °C, 200 rpm, amplitude: 30 mm, 5 % CO_2, aeration: A-D, BFT50, 10 ml cell suspension. Each point represents a mean of biological triplicates with resulting standard deviation (n = 3). Corresponding viable cell density at 100 %: 6.07·10^6 ± 0.08·10^6 cells·ml^{-1} (EF) / 6.17·10^6 ± 0.20·10^6 cells·ml^{-1} (IS). Corresponding hIgG concentration (titer) at 100 %: 291.0 ± 19.8 mg·l^{-1} (EF) / 178.0 ± 21.7 mg·l^{-1} (IS) at a viabilities ≥85 %.

4.3.3.4 Single cell cloning of K20-3/*cgr*Snord78

The polyclonal K20-3/*cgr*Snord78 cell line was single cell cloned (Chapter 3.8.4) to gain monoclonal cell lines for observing the clonal variations. In total sixteen clones derived from a single cell could be isolated. These clones were compared regarding their hIgG-productivity by hIgG-specific ELISA (Figure 4.32) as well as SDS-PAGE in a semi-quantitatively manner (data not shown). Two clones (K20-3/*cgr*Snord78/5E8 and K20-3/*cgr*Snord78/1D8) were selected for further comparison in an industrial relevant fedbatch process (Chapter 4.3.3.5).

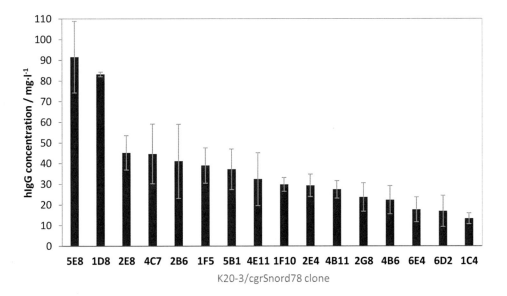

Figure 4.32:   ELISA hIgG-production data of K20-3/*cgr*Snord78 single cell clones. Each bar represents technical triplicates with the resulting standard deviation ($n_{biological}$ = 1). Medium: CM1035 + 5 mM Gln + 200 µg·ml^{-1} hygromycin B + 600 µg·ml^{-1} G418. Culture conditions: 37 °C, 5 % CO_2, 6-well, 2 – 3 ml cell suspension. Sampling was performed at day 6 of culture.

4.3.3.5 Application of K20-3/*cgr*Snord78 variants in a production fedbatch process

For final comparison, an optimised fedbatch process for hIgG-production in K20-3 cell line was applied on the stably transfected K20-3/*cgr*Snord78 polyclonal cell line as well as the derived clonal cell lines. This fedbatch process needed previous direct medium adaption of the cell lines to CDM4PerMAb + 4 mM Gln including selection antibiotics (Table 3.1). This procedure took several passages allowing robust cell growth.

At first, the Mock control cell line was compared with the freshly thawed and stably passaged progenitor cell line K20-3 to approve the applied Mock control (Figure 4.33).

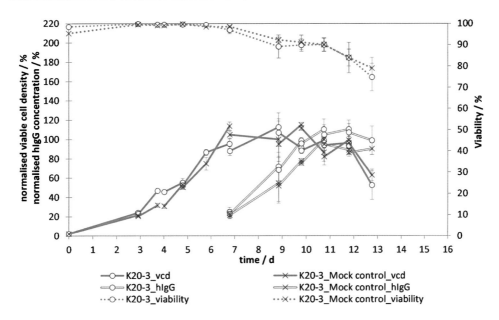

Figure 4.33: Comparision of K20-3 and K20-3/Mock control cell line. Medium: CDM4PerMAb + 4 mM Gln. Applied feeds: 50 g·l⁻¹ CellBoost 6 or 100 g·l⁻¹ glucose. Feeding of CellBoost6 (+ 2.5 g·l⁻¹) at day 3, 5, 7, 9 and 11. Feeding of CellBoost 6 (+ 3.0 g·l⁻¹) at day 4, 7 and 10. Culture conditions: 37 °C, 200 rpm, amplitude: 30 mm, 5 % CO_2, aeration: A-D, BFT50, 10 ml cell suspension. Each point represents a mean of biological triplicates with resulting standard deviation (n = 3). Corresponding viable cell density at 100 %: $13.84·10^6 ± 2.61·10^6$ cells·ml⁻¹. Corresponding hIgG concentration (titer) at 100 %: 912.5 ± 107.8 mg·l⁻¹.

Here, the mock control (K20-3/CV121lin) showed an identical growth and viability pattern compared to K20-3 cell line, but insignificant reduction of hIgG production as well. However, both cell lines were comparable and therefore the Mock control cell line was suitable for further comparisons.

Applying the optimised fedbatch procedure, the cells reached higher cell densities as well as hIgG titer in general. Nevertheless, K20-3/*cgr*Snord78 again reached similar and insignificant different cell densities as the K20-3/Mock control (Figure 4.34). Apart from to the previous fedbatch (Figure 4.31), the K20-3/*cgr*Snord78 cell line could not prolong the viable cell culture (≥ 85 % viability) under optimised conditions suggesting reduced limitation effects.

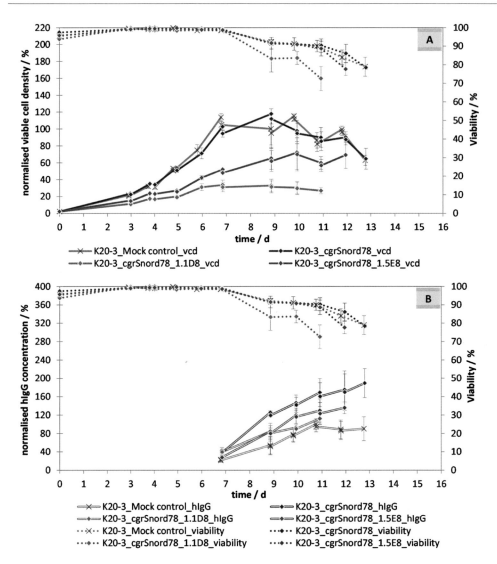

Figure 4.34: Production fedbatch on K20-3/*cgr*Snord78 cell line and its clonal variants in comparison with Mock cell line (K20-3/CV121lin). Normalised viable cell densities and hIgG concentrations are shown in A and B respectively. Medium: CDM4PerMAb + 4 mM Gln. Applied feeds: 50 g·l⁻¹ CellBoost 6 or 100 g·l⁻¹ glucose. Feeding of CellBoost 6 (+ 2.5 g·l⁻¹) at day 3, 5, 7, 9 and 11. Feeding of glucose (+ 3.0 g·l⁻¹) at day 4, 7 and 10. Culture conditions: 37 °C, 200 rpm, amplitude: 30 mm, 5 % CO_2, aeration: A-D, BFT50, 10 ml cell suspension. Each point represents a mean of biological triplicates with resulting standard deviation (n = 3). Corresponding viable cell density at 100 %: $13.84 \cdot 10^6 \pm 2.61 \cdot 10^6$ cells·ml⁻¹. Corresponding hIgG concentration (titer) at 100 %: 912.5 ± 107.8 mg·l⁻¹.

Regarding the hIgG concentration, the K20-3/*cgr*Snord78 pool cell line generated 2.0-fold more recombinant protein than the Mock control at ≥ 80 % viability (Figure 4.34 B). Therefore and due to the insignificant distinct integral cell density, the corresponding cell specific productivity increased 2.0-fold as well by sole overexpression of intronic cgrSnord78.

Furthermore, the clonal variants of K20-3/cgrSnord78 showed even higher increase in cell specific productivities (3.9-fold or 30.2 pg·cell^{-1}·d^{-1} for clone 1D8 and 2.8-fold or 21.7 pg·cell^{-1}·d^{-1} for clone 5E8, respectively). Nevertheless, the K20-3/cgrSnord78 clone 1D8 achieved no increase in hIgG concentration (1.0-fold, ≥ 80 % viability), whereat clone 5E8 increased production to 1.5-fold (Figure 4.34). Consequently, the K20-3/cgrSnord78 clone 1D8 produced 27 % of mock control integral cell density (≥ 80 % viability) only and was not able to maintain viable cell culture after day 10. On the other hand, the K20-3/cgrSnord78 clone 5E8 was able to achieve higher hIgG concentrations due to higher cell mass formation. Nevertheless, the cells preferentially produced the recombinant antibody instead of growing. This unbalanced production/growth relationship led to an insignificant change of hIgG concentration.

Therefore, under given points of view, the isolated K20-3/cgrSnord78 clones were not suitable for further hIgG production. Moreover, both clonal cell lines were not capable to produce more recombinant protein than the pool cell line. Therefore, a quantitative real time PCR approach was applied to investigate possible mechanisms underlying these significant differences in growth and production.

Here, the cgrSnord78 overexpression was constantly high in all clonal and polyclonal K20-3 derived cell lines time independently. Compared to the expression in the Mock cell line, the polyclonal K20-3/cgrSnord78 showed the highest grade of overexpression with 40.0-fold increase at day 4 and 58.1-fold increase at day 7, respectively (Figure 4.35). A similar expression pattern showed the clonal cell line K20-3/cgrSnord78/5E8 but with lower extent compared to the polyclonal cell line (30.8-fold overexpression at day 4 and 45.7-fold at day 7). The clone 1D8 showed an unchanged expression at day day 4 and 7 (both 30.8-fold overexpression compared to Mock cell line). Since the cell specific productivity of the isolated clonal cell lines was higher, the maturation/splicing rate of cgrSnord78, the change of intrinsic cgrGas5 expression as well as of the hIgG light (LC) and heavy chain (HC) transcription were determined (Figure 4.35). The splicing ability of cgrSnord78 intron was measured in comparison to the IntronA of Mock control. Here, the rate of splicing of cgrSnord78 intron was lower than for IntronA resulting an IntronA/cgrSnord78 ratio of at least 1.5-fold (clone 1D8, day 4) and 5.4-fold (clone 5E8, day 7) (Figure 4.35). Additionally, the ratios at day 7 were higher in all K20-3/cgrSnord78 cell lines either based on reduced splicing ability of cgrSnord78 intron compared to IntronA or increased IntronA splicing.

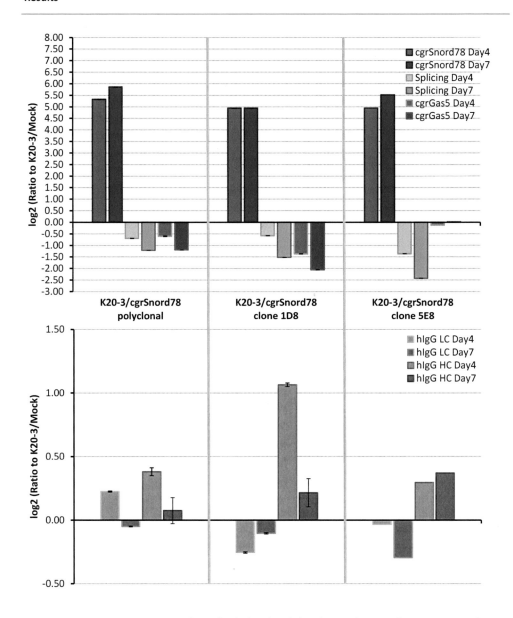

Figure 4.35: Quantitative RT-PCR data of polyclonal and clonal K20-3/*cgr*Snord78 variants. Medium: CDM4PerMAb + 4 mM Gln. Applied feeds: 50 g·l⁻¹ CellBoost 6 or 100 g·l⁻¹ glucose. Feeding of CellBoost6 (+ 2.5 g·l⁻¹) at day 3 and 5. Feeding of glucose (+ 3.0 g·l⁻¹) at day 4. Culture conditions: 37 °C, 200 rpm, amplitude: 30 mm, 5 % CO_2, aeration: A-D, BFT50, 10 ml cell suspension. Each bar represents technical duplicates of pooled biological triplicates with the resulting standard deviation. Sampling was performed at day 4 and day 7 respectively of culture in figure 4.34.

Therefore, no relationship between *cgr*Snord78 transcription/splicing and hIgG productivity could be observed. Basically, the content of putatively active *cgr*Snord78 (spliced intron, polyclonal > clone 1D8 > clone 5E8) was not comparable with cell specific productivity (clone 1D8 > clone 5E8 > polyclonal) or the integral viable cell density (polyclonal > clone 5E8 > clone 1D8).

The intrinsic transcription of *cgr*Gas5 was changed in case of the polyclonal and clonal 1D8 cell line compared to Mock control cell line. In K20-3/*cgr*Snord78 clone 5E8 the *cgr*Gas5 transcript content at day 4 and 7 was comparable to Mock cell line. Again, no correlation could be found between intrinsic *cgr*Gas5 transcript content and productivity or viability. Nevertheless, if the poor viability was connected to the strongest reduction of *cgr*Gas5 transcription in K20-3/*cgr*Snord78 clone 1D8 (2.6-fold at day 4 and 4.2-fold-reduction at day 7), this effect might be secondary due to overall low *cgr*Gas5 content ($1.01 \cdot 10^5$ copies of ActB compared to one copy of *cgr*Gas5 based on overall means for $Ct_{cgrGas5}$ of 31.0 ± 1.1 and mean Ct_{ActB} to 14.4 ± 0.3, n = 32).

Therefore, the transcription of hIgG fragments was determined. Here, the solely significant change in hIgG LC/HC transcript formation was observed at day 4 of K20-3/*cgr*Snord78 clone 1D8 cell culture (Figures 4.34 and 4.35) with a 2.1-fold upregulation of heavy chain transcription. On the other hand, at day 7 the upregulation was abolished and nearly levelled to Mock control transcription. Therefore, no linear transcription/productivity causality could be observed with the present parameters. Hence, other factors might interplay with *cgr*Snord78 and its maturation, so *cgr*Snord78 would rather increase the translation independently to the mRNA content of hIgG subunits (Chapter 5.4.2).

In summary, the increase in productivity could be proven by overexpression of intronic *cgr*Snord78. As an overexpressing polyclonal cell line (25-CHO-S/2.5C8 or K20-3) without present, spontaneous or experimentally inserted clonal variations, intronic *cgr*Snord78 was not affecting cell growth, maximal cell density or cell aggregation (data not shown) regardless the culture process (batch or fedbatch).

4.3.4 Validation of *cgr*Ttc36

In following, the isolation and validation of *C. griseus* Ttc36 (tetratricopeptide repeat domain 36) (*cgr*Ttc36) is described in detail.

4.3.4.1 Amplification of *cgr*Ttc36

Using the homologues sequences of Ttc36/TTC36 (*hsa*TTC36/HBP21/gi|122937278; *mmu*Ttc36/gi|20336733; *rno*Ttc36/gi|53850619) the respective primer pair (Table 3.14) was generated for amplification of the hamster Ttc36 (*cgr*Ttc36) sequence from CHO-S-E400/1.104 cDNA (Chapter 3.6.2). The resulting protein slightly differed from the recently uploaded hypothetical *cgr*Ttc36 sequence (Ttc36-like, gi|354496897) (Figure 4.36).

```
cgrTtc36_exp    MGTPNDQAVLQAIFNPNTPFGDVIDLDL-EEAKKED--EDGVFPQEQLEQSKALELQGVRAAE
cgrTtc36-like   MGTPNDQAVLQAIFNPNTPFGDVIDLDL-EEAKKED--EDGVFPQEQLEQSKALELQGVRAAE
mmuTtc36        MGTPNDQAVLQAIFNPDTPFGDVVGLDL-EEAEEGD--EDGVFPQAQLEHSKALELQGVRAAE
rnoTtc36        MGTPNDQAVLQAILNPNTPFGDVVGLDL-EETEEGD--EDGVFPQAQLEQSKALELQGVRAAE
hsaTTC36        MGTPNDQAVLQAIFNPDTPFGDIVGLDLGEEAEKEEREEDEVFPQAQLEQSKALELQGVMAAE
                *************:**:*****::.*** **::: :   ** **** ***:********* ***

cgrTtc36_exp    AGDLHTALEKFGQAICLLPERASAYNNRAQARRLQGDVAGALEDLERAVTLSGGQGRAARQSF
cgrTtc36-like   AGDLHTALEKFGQAICLLPERASAYNNRAQARRLQGDVAGALEDLERAVTLSGGQGRAARQSF
mmuTtc36        AGDLHTALEKFGQAISLLPDRASAYNNRAQARRLQGDVAGALEDLERAVTLSGGRGRAARQSF
rnoTtc36        AGDLHTALERFGKAISLLPERASAYNNRAQARRLQGDVAGALEDLERAVTLSGGRGRAARQSF
hsaTTC36        AGDLSTALERFGQAICLLPERASAYNNRAQARRLQGDVAGALEDLERAVELSGGRGRAARQSF
                **** ****:**:**.***:*************************** ****:********

cgrTtc36_exp    VQRGLLARLQGRDDDARRDFEQAARLGSPFARRQLVLLNPYAALCNRMLADMMGQLRAPSNGR
cgrTtc36-like   VQRGLLARLQGRDDDARRDFEQAARLGSPFARRQLVLLNPYAALCNRMLADMMGQLRAPSNGH
mmuTtc36        VQSGLLARFQGRDDDARRDFEKAARLGSPFARRQLVLLNPYAALCNRMLADMMGQLRAPSNGR
rnoTtc36        VQRGLLARFQGRDDDARRDFEQAARLGSSFARRQLVLLNPYAALCNRMLADMMGQLSAPSNGR
hsaTTC36        VQRGLLARLQGRDDDARRDFERAARLGSPFARRQLVLLNPYAALCNRMLADMMGQLRRPRDSR
                ** *****:***********:.******.************************* * :.:
```

Figure 4.36: Protein sequence alignment of the amplified sequence with the documented Ttc36/TTC36 species (*cgr*Ttc36_exp: experimental *cgr*Ttc36 sequence amplified; putative *cgr*Ttc36-like: XP_003510560.1 / gi|354496897; *mmu*Ttc36: NP_620401.1 / gi|20336733; *rno*Ttc36: NP_001005546.1 / gi|53850619; *hsa*TTC36: NP_001073910.1 / gi|122937278). Consensus sequences are back grounded in grey and the three tetratricopeptide repeat domains (after Liu *et al.* 2008) are highlighted in shades of red.

Nevertheless, the isoelectric point (pI = 5.29) of amplified and known sequence did not differ. In addition, the crucial tetratricopeptide repeat domains were shown to be highly homologous through all regarded species unlike the N-terminal as well as the far C-terminal residues.

Before transfection (Chapter 3.8.2), the amplified *cgr*Ttc36 was inserted within the CV121 vector construct after chapter 3.8.1. Subsequently, *cgr*Ttc36 was overexpressed in non-producer as well as producer cell lines for effect validation (Chapter 4.3.4.2).

4.3.4.2 Validation by overexpression in non-producing cell line 25-CHO-S/2.5C8

First, the clonal progenitor cell line 25-CHO-S/2.5C8 was transfected with linearised CV146 (CV146lin) vector (Chapter 3.7.12) for observation of the change in cell density and overall growth as well as viability behaviour. Surprisingly, the overexpression of *cgr*Ttc36 led to an increase in viable cell density (1.78-fold and 1.88-fold, respectively) during identical fedbatch cultures (Figure 4.37). Interestingly, the grade of increase in viable cell density was similar to the EF-fed fedbatch cultures of mutant cell line CHO-S-E400/1.104 against progenitor cell line 25-CHO-S/2.5C8 (Chapter 4.1.9, Figure 4.16). That led to the suggestion that *cgr*Ttc36 was a major regulator of viable cell density increase in CHO-S-E400/1.104. Furthermore, overexpression of *cgr*Ttc36 prolonged the viable cell culture ($\geq$ 80% viability) for at least one day (Figure 4.37).

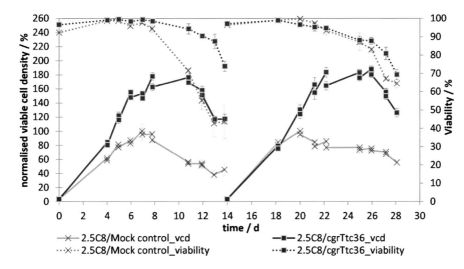

Figure 4.37: Overexpression of the *cgr*Ttc36 in 25-CHO-S/2.5C8 and comparison with Mock cell line (25-CHO-S/2.5C8/CV121lin) in parallel as well as serial fedbatch cultures. Medium: 87.5 % CM1035 / 10 % CHO CD EfficientFeed A/B + 5 mM Gln + 200 µg·ml⁻¹ hygromycin B. Feed: 80 % CHO CD EfficientFeed A/B + 40 mM Gln + 200 µg·ml⁻¹ hygromycin B. Feeding rate: 500 µl at day 4, 5, 6, 7, 11, 12, 13, 18, 20, 25, 26 and 27; 750 µl at day 21; 1000 µl at day 8; 1250 µl at day 22. Culture conditions: 37 °C, 200 rpm, amplitude: 30 mm, 5 % CO_2, aeration: A-D, BFT50, 10 ml cell suspension. Each point represents a mean of biological triplicates with resulting standard deviation (n = 3). Corresponding viable cell density at 100 %: 7.40·10⁶ cells·ml⁻¹ (1st fedbatch culture) and 7.50·10⁶ cells·ml⁻¹ (2nd fedbatch culture), respectively (Maximal viable cell density of mock control).

During continuous culture, the cell line 25-CHO-S/2.5C8/*cgr*Ttc36 could maintain the ability to growth at high cell densities (monitored for at least 20 passages). But interestingly, after cryopreservation and thawing circles (Chapter 3.4) the effect was diminished or even abolished under constantly high *cgr*Ttc36 mRNA transcription (data not shown). Suggesting the methylation of CpG island enriched within the coding sequence of *cgr*Ttc36 (Figure 4.38), the polyclonal cell line 25-CHO-S/2.5C8/*cgr*Ttc36 was treated with a gradient of 0.5 μM to 5.0 μM 5-aza-2'-deoxycytidine. Here, the viable cell density could not be increased again.

```
ATGGGGACTCCAAATGATCAGGCAGTGCTGCAGGCCATCTTCAACCCCAACACACCATTTGGAGATGTCATTGACTTGG
ACCTGGAAGAAGCAAAGAAAGAAGATGAAGATGGAGTTTTCCCTCAAGAACAGTTGGAGCAGTCCAAAGCTCTGGAGTT
GCAGGGAGTGAGGGCAGCAGAAGCTGGGGACCTCCACACAGCCCTGGAGAAGTTTGGCCAAGCTATCTGCCTGCTACCT
GAGAGAGCCTCTGCCTACAACAACCGGGCTCAAGCCCGGAGGCTCCAGGGGGATGTAGCAGGCGCCCTGGAGGACTTGG
AGCGCGCAGTGACGCTGAGCGGCGGCCAGGGTCGCGCCGCCCGCCAGAGCTTCGTGCAGCGCGGACTGCTGGCGCGATT
GCAAGGCGGAGACGACGACGCCCGCAGGGACTTCGAGCAGGCAGCGCGACTGGGCAGCCCGTTCGCGCGGCGCCAGCTG
GTGCTGCTCAACCCGGTACGCCGCGCGCTGTGCAACCGCATGCTGGCCGGACATGATGGGGCAGCTACGCGCGCCCAGTAACG
GGCGGCTGA
```

Figure 4.38: The coding sequence of amplified and introduced *cgr*Ttc36 with special consideration on in CpG island content. The CpG Islands (shaded in red, white letters) are enriched within the 3'-terminus of the *cgr*Ttc36 coding sequence.

Suggesting a C→T transitions due to CpG methylation (Walser and Furano 2010; Misawa and Kikuno 2009; Jiang and Zhao 2006), the *cgr*Ttc36 coding sequence of the thawed 25-CHO-S/2.5C8/*cgr*Ttc36 cell population was amplified and sequenced subsequently to gDNA isolation (Chapter 3.7.1). Putative C→T transitions could lead to severe functional loss or truncation of *cgr*Ttc36 protein. Nevertheless, the sequence was intact suggesting another unknown effect inhibition (direct negative feedback or a methylation-independent epigenetic modulation of *cgr*Tc36 or antagonists/agonists) (data not shown).

Therefore, *cgr*Ttc36 was subsequently validated in hIgG-producing cell line K20-3 for reproducing the increase in viable cell density and determine its effect on productivity (Chapter 4.3.4.3).

4.3.4.3 Further effect validation by overexpression in hIgG-producing cell line K20-3

As the overexpression in 25-CHO-S/2.5C8 revealed a significant increase in maximal cell density as well as integrated viability, particular attention was given to the influence on the hIgG production by overexpression in K20-3. Since hIgG$_A$ (other specificity than the K20-3 produced antibody) expression in transfected mutant CHO-S-E400/1.104 cell line was approx. 2-fold lower than in parallel treated control cell line (Chapter 4.1.7), it was elucidated whether the increase in cell density and growth by overexpression of *cgr*Ttc36 would lead to a reduced productivity. Hence, the linearised CV146lin vector (Chapter 3.8.1) was stably transfected in K20-3 producer cell line, which was initially compared with the mock transfected cell line in batch culture (Figure 4.39).

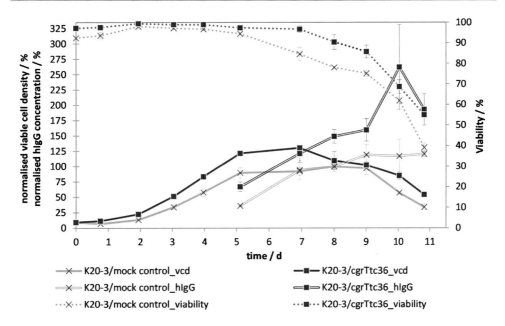

Figure 4.39: Overexpression of *cgr*Ttc36 in mAb-producing K20-3 cell line and comparison with Mock cell line (K20-3/CV121lin) in parallel batch cultures. Medium: CM1035 + 5 mM Gln + 200 µg·ml⁻¹ hygromycin B + 600 µg·ml⁻¹ G418. Culture conditions: 37 °C, 200 rpm, amplitude: 30 mm, 5 % CO_2, aeration: A-D, BFT50, 10 ml cell suspension. Each point represents a mean of biological triplicates with resulting standard deviation (n = 3). Corresponding viable cell density at 100 %: $5.43 \cdot 10^6$ cells·ml⁻¹. Corresponding mean hIgG concentration (titer) at 100 %: 146.8 mg·l⁻¹ at a viability of 78.0 ± 0.5 %.

Here, the maximal cell density was increased 1.30-fold compared to mock cell line (K20-3/CV121lin). In parallel, the hIgG titer was increased 1.69-fold regarding values with viabilities of ≥85 %. But due to the higher integrated cell density (2.12-fold, viabilities ≥85 %) the specific productivity was reduced about 20 %. As observed in figure 4.37, the viable cell culture was prolonged for at least one day compared to K20-3/mock control cell line.

The lower extent in viable cell density increase was suggested to base on the energy consuming recombinant hIgG production process, which was absent in 25-CHO-S/2.5C8/*cgr*c36. Therefore, the effect of *cgr*Ttc36 was elucidated by applying a fedbatch process to overcome possible limitation effects (Figure 4.40).

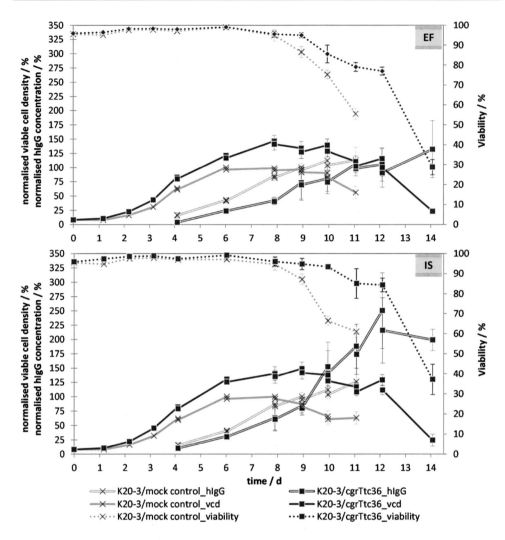

Figure 4.40: Overexpression of *cgr*Ttc36 in mAb-producing K20-3 cell line and comparison with Mock cell line (K20-3/CV121lin) in parallel fedbatch cultures using CHO CD EfficientFeed A/B (EF) or IS CHO Feed-CD XP (IS). Medium: CM1035 + 5 mM Gln + 200 μg·ml^{-1} hygromycin B + 600 μg·ml^{-1} G418. Applied feed (EF or IS) composition: 80 % Feed + 40 mM Gln + 200 μg·ml^{-1} hygromycin B + 600 μg·ml^{-1} G418. Feeding rate (corresponding to L-Glutamine end concentration): 1.5 mM at day 4, 6 and 8; 2.0 mM at day 9; 3.0 mM at days 10 and 11. Total feed (EF or IS) intake (cumulative): 22.0 % at day 11. Culture conditions: 37 °C, 200 rpm, amplitude: 30 mm, 5 % CO_2, aeration: A-D, BFT50, 10 ml cell suspension. Each point represents a mean of biological triplicates with resulting standard deviation (n = 3). Corresponding viable cell density at 100 %: 6.07·10^6 ± 0.08·10^6 cells·ml^{-1} (EF) / 6.17·10^6 ± 0.20·10^6 cells·ml^{-1} (IS). Corresponding hIgG concentration (titer) at 100 %: 291.0 ± 19.8 mg·l^{-1} (EF) / 178.0 ± 21.7 mg·l^{-1} (IS) at a viabilities ≥85 %.

Here, the two different feed media (CHO CD EfficientFeed A/B [EF] and IS CHO Feed-CD XP [IS]) were applied in parallel to observe the efficiencies of both media to putatively increase growth, the maximal cell density as well as the productivity. Both fedbatch culture processes were carried out using the same feed ratios and schedules (Figure 4.40).

Applying both feed media in parallel, no significant difference in maximal cell density increase was observed (1.46-fold for EF-fedbatch and 1.49-fold for IS-fedbatch cultures). The latter cell density increase was only 1.25-fold EF-fedbatch and 1.30-fold IS-fedbatch cultures compared to batch culture (Figure 4.39). In comparison, the viable cell density increase in mock control cell line fedbatch culture was increased in lower extent than in K20-3/*cgr*Ttc36 fedbatches (1.12-fold [EF] and 1.14-fold [IS]). Therefore, the K20-3/*cgr*Ttc36 cell line has undergone a limitation during batch culture in respect to the viable cell density.

The viable cell culture (≥ 75 % viability) was extended from three days by *cgr*Ttc36 overexpression in EF-fedbatches as well as IS-fedbatches, although cultures fed with IS CHO Feed-CD XP [IS] had a longer period above viabilities of 85 % (Figure 4.40).

The achieved volumetric productivities (hIgG titers) in K20-3/*cgr*Ttc36 fedbatch cultures above viabilities of 75 % were 233.5 ± 64.6 mg·l^{-1} (85.5 ± 4.5 % viability) or 306.2 ± 85.7 mg·l^{-1} (77.0 ± 2.0 % viability) for the EF-fedbatch and 445.5 ± 117.8 mg·l^{-1} (84.3 ± 3.4 % viability) for the IS-fedbatch cultures. As highlighted in chapter 4.3.3.3, the mock control titers were higher in EF-fedbatches than in IS-fedbatches. Therefore, the increase in hIgG titer in comparison to the mock control fedbatches was unequally distributed in K20-3/*cgr*Ttc36 fedbatch cultures. Due to this occurence, the increases in hIgG titers of K20-3/*cgr*Ttc36 fedbatch cultures (≥ 75 % viability) were 0.94-fold (EF) and 2.50-fold (IS) in maximum. In terms of mean cell specific productivity, in EF-fedbatch cultures (≥ 75 % viability) the K20-3/*cgr*Ttc36 cell line reached 4.6 pg·cell^{-1}·d^{-1} compared to mock control with 8.6 pg·cell^{-1}·d^{-1} and in IS-fedbatch cultures (≥ 85 % viability) the K20-3/*cgr*Ttc36 cell line reached 6.4 pg·cell^{-1}·d^{-1} compared to mock control with 5.4 pg·cell^{-1}·d^{-1}. Therefore, the mean cell specific productivities increased 0.53-fold (EF) and 1.16-fold respectively.

These data suggested a cell density linked hIgG-titer increase without affecting the cell specific productivity by *cgr*Ttc36 overexpression in K20-3 cell line. Again, the feed dependent productivity in mock control cell line (Chapter 4.3.3.3; Figure 4.31) affecting the comparison.

In addition to the results in productivity and cell density, overexpression of *cgr*Ttc36 in K20-3 significantly reduced cell aggregation (Figure 4.41). This observation covered the results of overexpression of *cgr*C10orf93-like (putative *cgr*Ttc40-like) (Figure 4.25). Therefore, it was suggested that the observed aggregation inhibition would underlay a common mechanism.

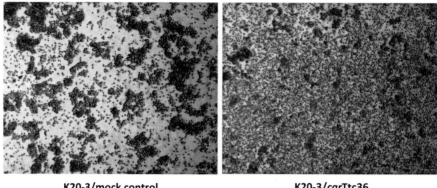

| K20-3/mock control | K20-3/*cgr*Ttc36 |

Figure 4.41: Overexpression of *cgr*Ttc36 in mAb-producing K20-3 cell line led to reduced cell aggregation compared to the mock control. Medium: 87.5 % CM1035 / 10 % CHO CD EfficientFeed A/B + 5 mM Gln + 200 µg·ml^{-1} hygromycin B. Culture conditions: 37 °C, 5 % CO_2, 6-well plate, 2 ml cell suspension. The pictures are representative for the respective cell line and were generated six days after seeding.

For further elucidation of clonal variations in maximal cell density, productivity as well as aggregation behaviour, single cell clones of the K20-3/*cgr*Ttc36 polyclonal cell line were generated (Chapter 4.3.4.4).

4.3.4.4   Clonal variations in K20-3/*cgr*Ttc36

For a next step, the polyclonal cell line K20-3/*cgr*Ttc36 was singled and separated into monoclonal cell populations by limited dilution (Chapter 3.8.4). The applied procedure (Table 3.22) resulted in six clonal cell lines (1.5F5, 1.7C1, 1,7E4, 1.7G8, 1.12A8 and 1.17D7) with different growth (Figure 4.42) and hIgG production characteristics (Figure 4.43).

The clonal K20-3/*cgr*Ttc36 cell lines possessed similar growth characteristics, but clone 1.12A8 with overall poor growth rate, low cell density and shortest viable culture duration of all clones (Figure 4.42). On the other hand, the clones 1.7C1 as well as 1.17D7 possessed higher initial growth rates than the other clonal cell lines and together with clone 1.7G8 the longest viable culture duration (≥80 % viability) of at least one day prolongation compared to K20-3/*cgr*Ttc36 pool cell line. The clones 1.5F5 and 1.7E4 showed a shortened viable culture duration around one day instead.

Interestingly, the K20-3/*cgr*Ttc36 clone 1.5F5 revealed an 1.75-fold increase in volumetric hIgG productivity (≥80 % viability) (Figure 4.43). All other K20-3/*cgr*Ttc36 clones and the pool cell line showed similar volumetric productivities. In addition, the K20-3/*cgr*Ttc36 clone 1.5F5 possessed a 2.20-fold increase in cell specific productivity, which was similar to clone 1.12A8 (Figure 4.44).

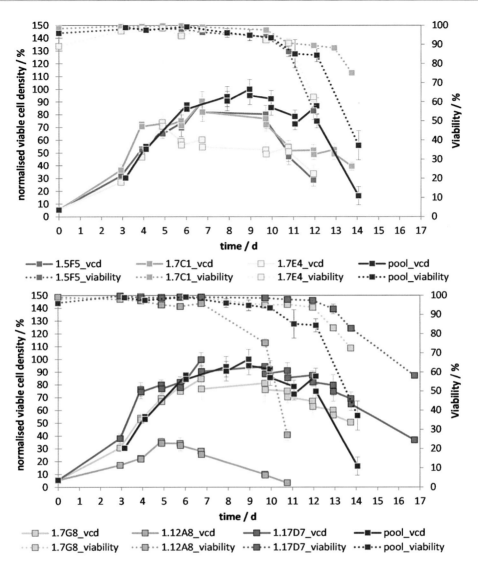

Figure 4.42: Viable cell density characteristics of isolated K20-3/*cgr*Ttc36 single cell clones in parallel fedbatch cultures. Medium: CM1035 + 5 mM Gln + 200 µg·ml^{-1} hygromycin B + 600 µg·ml^{-1} G418. Applied feed composition: 80 % IS CHO Feed-CD XP + 40 mM Gln + 200 µg·ml^{-1} hygromycin B + 600 µg·ml^{-1} G418. Feeding rate (corresponding to L-Glutamine end concentration): +1.0 mM at day 4; +1.5 mM at day 5; +2.0 mM at day 6; +3.75 mM at day 7; +2.5 mM at days 10, 11, 12, 13 and 14. Total feed intake (cumulative): 33.2 % at day 14. Culture conditions: 37 °C, 200 rpm, amplitude: 30 mm, 5 % CO_2, aeration: A-D, BFT50, 10 ml cell suspension. Each point represents a mean of biological triplicates with resulting standard deviation (n = 3). Corresponding viable cell density at 100 %: 9.60·10^6 ± 0.70·10^6 cells·ml^{-1}. Pool K20-3/*cgr*Ttc36 culture values correspond with data in figure 4.40.

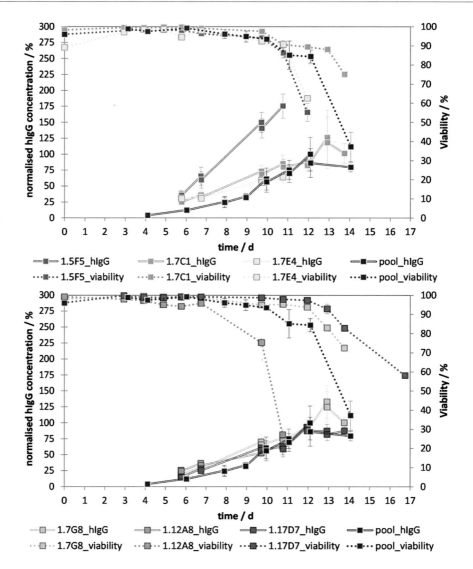

Figure 4.43: Product formation characteristics of isolated K20-3/*cgr*Ttc36 single cell clones in parallel fedbatch cultures. Medium: CM1035 + 5 mM Gln + 200 µg·ml^{-1} hygromycin B + 600 µg·ml^{-1} G418. Applied feed composition: 80 % IS CHO Feed-CD XP + 40 mM Gln + 200 µg·ml^{-1} hygromycin B + 600 µg·ml^{-1} G418. Feeding rate (corresponding to L-Glutamine end concentration): +1.0 mM at day 4; +1.5 mM at day 5; +2.0 mM at day 6; +3.75 mM at day 7; +2.5 mM at days 10, 11, 12, 13 and 14. Total feed intake (cumulative): 33.2 % at day 14. Culture conditions: 37 °C, 200 rpm, amplitude: 30 mm, 5 % CO_2, aeration: A-D, BFT50, 10 ml cell suspension. Each point represents a mean of biological triplicates with resulting standard deviation (n = 3). Corresponding hIgG concentration at 100 %: 445.5 ± 117.8 mg·l^{-1}. Pool K20-3/*cgr*Ttc36 culture values correspond with data in figure 4.40.

In contrast to clone 1.5F5, the K20-3/*cgr*Ttc36 clone 1.12A8 possessed a similar cell specific productivity increase (2.07-fold, ≥75 % viability) compared to pool cell line, but without a balanced cell growth the hIgG titer maintained on the level of the K20-3/*cgr*Ttc36 pool cell line. The clonal cell lines 1.5F5 and 1.12A8 showed cell specific productivities between 8 – 14 pg·cell^{-1}·d^{-1} instead of 2 – 8 pg·cell^{-1}·d^{-1} for the other single cell clones as well as the pool cell line (Figure 4.44). Furthermore, it could be shown that the cell specific productivities increased during culture and prolonged culture duration. In summary, the K20-3/*cgr*Ttc36 clone 1.5F5 showed the best characteristics for producing the present hIgG.

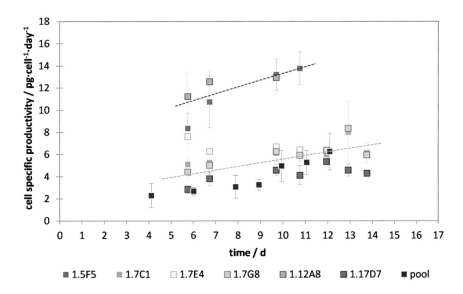

Figure 4.44:   Cell specific productivities of isolated K20-3/*cgr*Ttc36 single cell clones during parallel fedbatch cultures. Trend lines of each cell specific productivity subgroup are dashed. Medium: CM1035 + 5 mM Gln + 200 µg·ml^{-1} hygromycin B + 600 µg·ml^{-1} G418. Applied feed composition: 80 % IS CHO Feed-CD XP + 40 mM Gln + 200 µg·ml^{-1} hygromycin B + 600 µg·ml^{-1} G418. Feeding rate (corresponding to L-Glutamine end concentration): +1.0 mM at day 4; +1.5 mM at day 5; +2.0 mM at day 6; +3.75 mM at day 7; +2.5 mM at days 10, 11, 12, 13 and 14. Total feed intake (cumulative): 33.2 % at day 14. Culture conditions: 37 °C, 200 rpm, amplitude: 30 mm, 5 % CO_2, aeration: A-D, BFT50, 10 ml cell suspension. Each point represents a mean of biological triplicates with resulting standard deviation (n = 3). Pool K20-3/*cgr*Ttc36 culture values correspond with data in figure 4.40.

Another variant process relevant parameter was the ability for aggregation (Figure 4.45). Here, the K20-3/*cgr*Ttc36 clones 1.12A8 possessed the highest grade of aggregation (approx. 75 % aggregated cell ratio) followed by 1.7E4 and 1.7G8 (each 40-60 % aggregated cell ratio). In addition, the high-producing clonal cell line K20-3/*cgr*Ttc36/1.5F5 showed a slightly enhanced cell aggregation ability compared to the K20-3/*cgr*Ttc36 pool cell line as well.

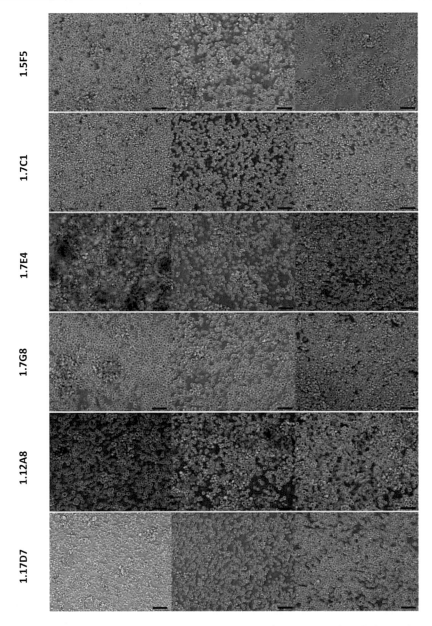

1.5F5

1.7C1

1.7E4

1.7G8

1.12A8

1.17D7

Figure 4.45: Aggregation characteristics of isolated K20-3/*cgr*Ttc36 single cell clones during parallel cultures. Samples were taken at day 4 before feed addition and pictures represent each replicate (n = 3) representatively. Medium: CM1035 + 5 mM Gln + 200 $\mu g \cdot ml^{-1}$ hygromycin B + 600 $\mu g \cdot ml^{-1}$ G418. Culture conditions: 37 °C, 200 rpm, amplitude: 30 mm, 5 % CO_2, aeration: A-D, BFT50, 10 ml cell suspension. The bars represent 100 μm.

By determination of the transcription level of total *cgr*Ttc36 mRNA (Figure 4.46), it was shown that *cgr*Ttc36 was not overexpressed in every clonal cell line. In single cell clone 1.12A8 the transcription of *cgr*Ttc36 was even slightly suppressed compared to the mock cell line (0.82-fold). The *cgr*Ttc36 mRNA level of the K20-3/*cgr*Ttc36 clones 1.7E4 (1.12-fold) and 1.7G8 (1.21-fold) was not significantly changed compared to the mock control. The slight downregulation of *cgr*Ttc36 in 1.12A8 would fit to the poor growth (Figure 4.42) as well as aggregation characteristics (Figure 4.45). Additionally, the higher grade aggregating clones 1.7E4 and 1.7G8 showed lower *cgr*Ttc36 expression as well compared to the K20-3/*cgr*Ttc36 pool cell line (62.9-fold) and the low aggregating single cell clones 1.5F5 (15.7-fold) , 1.7C1 (60.6-fold) and 1.17C7 (1.94-fold).

Taken together, first coincidences regarding growth as well as aggregation characteristics with *cgr*Ttc36 mRNA level could be shown. Nevertheless, the full connection between all these parameters needs to be elucidated.

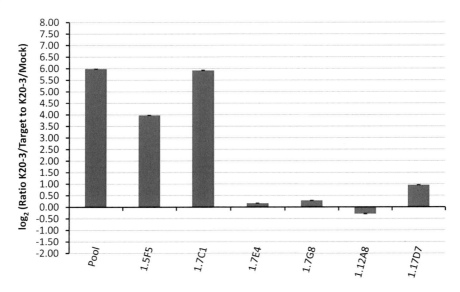

Figure 4.46: Quantitative real-time PCR data of K20-3/*cgr*Ttc36 polyclonal cell line and derived single cell clones. Each bar represents technical duplicates of pooled biological triplicates with the resulting standard deviation. Medium: CM1035 + 5 mM Gln + 200 μg·ml^{-1} hygromycin B + 600 μg·ml^{-1} G418. Culture conditions: 37 °C, 200 rpm, amplitude: 30 mm, 5 % CO_2, aeration: A-D, BFT50, 10 ml cell suspension. Sampling was performed at day 4 of culture during late exponential phase (single cell clone: Figure 4.42; pool cell line: Figure 4.40).

Further, the K20-3/*cgr*Ttc36 polyclonal cell line as well as 1.5F5 clonal cell line should be compared with the mock cell line within a production fedbatch process to prove their use in a biopharmaceutical production process (Chapter 4.3.4.5).

4.3.4.5   Application of K20-3/*cgr*Ttc36 variants in a production fedbatch process

For final comparison, an optimised fedbatch process for hIgG-production in K20-3 cell line was applied on the stably transfected K20-3/*cgr*Ttc36 polyclonal cell line as well as its derived 1.5F5 clone. As performed in parallel to chapter 4.3.3.5, the cell lines were adapted to CDM4PerMAb + 4 mM Gln including selection antibiotics (Table 3.1) by several passages in prior allowing robust cell growth.

As shown in chapter 4.3.3.5, the optimised fedbatch procedure resulted in higher cell densities as well as hIgG titer in general (Figure 4.47). Further, the polyclonal K20-3/*cgr*Ttc36 pool cell line reached significant higher integral viable cell densities (1.52-fold) as well as a higher maximal cell density (1.30-fold) than the K20-3/Mock control cell line ($\geq$ 80 % viability) (Figure 4.47 A). Comparable to the previous batch as well as fedbatches (Figures 4.39 and 4.40), the viable cell culture of the K20-3/*cgr*Ttc36 pool cell line ($\geq$ 80 % viability) was prolonged for two days under optimised conditions (for viabilities $\geq$ 85 %: three days). Under identical culture conditions, the viability characteristics of the clonal cell line K20-3/*cgr*Ttc36/1.5F5 were comparable with the pool cell line. Unlike the pool cell line, the maximal cell density of the clone 1.5F5 slightly as well as insignificantly increased (1.06-fold) compared to the mock cell line K20-3/CV121lin only. The integral viable cell density increased 1.25-fold due to enhanced viability characteristics.

Regarding the hIgG concentration, the volumetric productivity of the K20-3/*cgr*Ttc36 pool cell line increased 1.32-fold compared to the mock cell line (at $\geq$ 80 % viability) (Figure 4.47 B). On the other hand, the hIgG titer increased 3.26-fold in K20-3/*cgr*Ttc36/1.5F5 fedbatches and at viabilities $\geq$ 80 %. This resulted in a 2.60-fold increase (to 20.2 $\pm$ 3.8 pg·cell^{-1}·d^{-1}) in cell specific productivity compared to mock control cell line and a 2.98-fold increase compared to K20-3/*cgr*Ttc36 pool cell line even at viabilities $\geq$ 85 %. Therefore, the cell specific productivity of the pool cell line resulted in 1.15-fold decrease of the cell specific productivity at viabilities $\geq$ 80 % compared to mock cell line.

Taken together, the statistical overexpression of *cgr*Ttc36 in pool cell line without manifesting clonal divergences led to improved growth and viability characteristics and reduced cell specific productivities (Figures 4.40 and 4.47). Nevertheless, the desired increase in hIgG titer was achieved in combination with improved downstream processing by reduced cell aggregation (data were comparable for all feed media applied). Using another feed regime, the productivities could be increased, since at day 10 of the process relevant fedbatch (Figure 4.47 B) cell density was gradually decreased to presumably maintain the high viability without producing further recombinant protein. This would suggest a special nutrient deprivation after day 10. Furthermore, the clonal cell line K20-3/*cgr*Ttc36/1.5F5 achieved significantly higher hIgG concentrations (Figures 4.43 and 4.47 B) than K20-3/*cgr*Ttc36 pool as well as mock cell line. For subsequent elucidation of some molecular parameters, a quantitative real time PCR approach was applied to investigate possible mechanisms underlying these significant differences in growth and production of K20-3/*cgr*Ttc36 pool and K20-3/*cgr*Ttc36/1.5F5 clonal cell line (Figure 4.48).

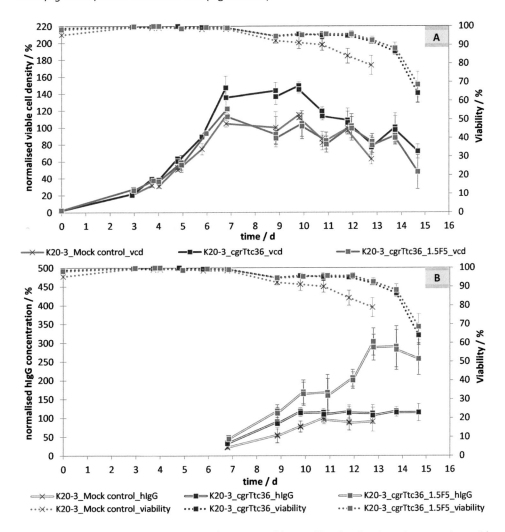

Figure 4.47: Production fedbatch on K20-3/*cgr*Ttc36 cell line and its clonal variants in comparison with Mock cell line (K20-3/CV121lin). Normalised viable cell densities and hIgG concentrations are shown in A and B respectively. Medium: CDM4PerMAb + 4 mM Gln. Applied feeds: 50 g·l⁻¹ CellBoost 6 or 100 g·l⁻¹ glucose. Feeding of CellBoost 6 (+ 2.5 g·l⁻¹) at day 3, 5, 7, 9, 11 and 13. Feeding of glucose (+ 3.0 g·l⁻¹) at day 4, 7, 10, 12 and 14. Culture conditions: 37 °C, 200 rpm, amplitude: 30 mm, 5 % CO_2, aeration: A-D, BFT50, 10 ml cell suspension. Each point represents a mean of biological triplicates with resulting standard deviation (n = 3). Corresponding viable cell density at 100 %: $13.84 \cdot 10^6 \pm 2.61 \cdot 10^6$ cells·ml⁻¹. Corresponding hIgG concentration (titer) at 100 %: 912.5 ± 107.8 mg·l⁻¹.

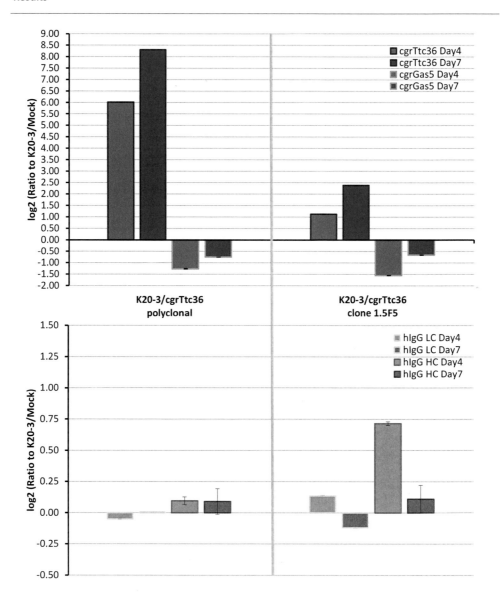

Figure 4.48: Quantitative RT-PCR data of polyclonal K20-3/*cgr*Ttc36 pool cell line and its clonal variant 1.5F5. Medium: CDM4PerMAb + 4 mM Gln. Applied feeds: 50 $g \cdot l^{-1}$ CellBoost 6 or 100 $g \cdot l^{-1}$ glucose. Feeding of CellBoost6 (+ 2.5 $g \cdot l^{-1}$) at day 3 and 5. Feeding of glucose (+ 3.0 $g \cdot l^{-1}$) at day 4. Culture conditions: 37 °C, 200 rpm, amplitude: 30 mm, 5 % CO_2, aeration: A-D, BFT50, 10 ml cell suspension. Each bar represents technical duplicates of pooled biological triplicates with the resulting standard deviation. Sampling was performed at day 4 and day 7 of culture in figure 4.47, respectively.

The *cgr*Ttc36 expression was shown to be similar to previous experiments (Figure 4.46) and in addition increased by time. Here, the K20-3/*cgr*Ttc36 pool cell line showed a transcription ratio of 64.7 ± 0.1 at day 4 and 315.2 ± 0.5 at day 7 respectively in comparison with mock cell line transcription. The transcription ratios for the clonal K20-3/*cgr*Ttc36/1.5F5 cell line were 2.18 ± 0.00 at day 4 and 5.21 ± 0.02 at day 7, respectively. Therefore, for K20-3/*cgr*Ttc36/1.5F5 the grade of overexpression was significantly lower in recent fedbatch culture (Figure 4.47) than in the previous approach (Figure 4.46). This discrepancy could give an explanation for the low maximal cell density of K20-3/*cgr*Ttc36/1.5F5 (Figures 4.42 and 4.47), which was similar to the mock cell line.

Interestingly, in both cell lines the *cgr*Gas5 transcription ratios to mock cell line were comparable and suggested to be not responsible for the observed differences in growth (Figures 4.47 and 4.48).

And finally, the transcription of hIgG low chain (LC) and heavy chain (HC) fragments did not significantly change (Figure 4.48). Therefore, mRNA stability, the translation or protein processing efficiency should be increased.

Therefore, under given points of view both the K20-3/*cgr*Ttc36 pool cell line and the isolated clone 1.5F5 were suitable for biopharmaceutical process application. Again, the pool cell line could not reach significant higher hIgG concentrations than the applied control cell line, but the overexpression of *cgr*Ttc36 could improve the overall growth characteristics. In addition, its clonal variant K20-3/*cgr*Ttc36/1.5F5 was able to improve the hIgG concentration by increased cell specific productivities and maintained viability characteristics suggesting a genomic integration site-specific effect.

4.3.4.6   An excursus: The co-expression of *cgr*Ttc36 with *cgr*Snord78 and *cgr*TPRp

Up to now, the factors were validated and elucidated in an isolated manner. In parallel, combinations of two factors were applied as well. Here, *cgr*Ttc36 (Chapter 4.3.4) was co-expressed with *cgr*TPRp (Chapter 4.3.1) or *cgr*Snord78 (Chapter 4.3.3.3) either. The expression vectors (CV147 and CV156 respectively; Table 3.21) were constructed and tested by stable transfection with the single targets in parallel. The primers were designed to provide multiple possibilities for combination.

For validation, the positively high cell density factor *cgr*Ttc36 was put downstream of *cgr*TPRp and an IRES sequence (Table 3.21) to couple their (over-)expression. On the other hand after establishing *cgr*Snord78, the target was coupled with *cgr*Ttc36 as well. Both combinations were transfected into the non-producing cell line 25-CHO-S/2.5C8. In fedbatches (Figure 4.49), both combinations could not show the desired effect of enhancing the maximal cell density. Solely, the combination *cgr*TPRp-IRES-*cgr*Ttc36 was slightly better than the separated overexpression of *cgr*TPRp.

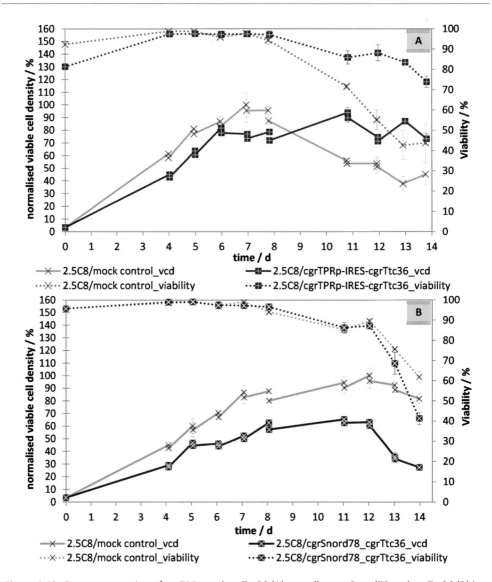

Figure 4.49: Co-overexpression of *cgr*TPRp and *cgr*Ttc36 (A) as well as *cgr*Snord78 and *cgr*Ttc36 (B) in 25-CHO-S/2.5C8 and comparison with Mock cell line (25-CHO-S/2.5C8/CV121lin) in parallel fedbatch cultures. Medium: 87.5 % CM1035 / 10 % CHO CD EfficientFeed A/B + 5 mM Gln + 200 µg·ml^{-1} hygromycin B. Feed: 80 % CHO CD EfficientFeed A/B + 40 mM Gln + 200 µg·ml^{-1} hygromycin B. Feeding rate: 500 µl at day 4, 5, 6, 7, 11 and 12 (B: and 13); 1000 µl at day 8. Culture conditions: 37 °C, 200 rpm, amplitude: 30 mm, 5 % CO_2, aeration: A-D, BFT50, 10 ml cell suspension. Each point represents a mean of biological triplicates with resulting standard deviation (n = 3). Corresponding viable cell density at 100 %: 7.40·10^6 cells·ml^{-1} (A) / 7.08·10^6 cells·ml^{-1} (B). (Maximal viable cell density of mock control).

Here (Figure 4.49 A), the viable cell culture duration (≥ 80 % viability) was prolonged for two days compared to the parallel fedbatch with 25-CHO-S/2.5C8/*cgr*TPRp (Figure 4.21). The growth rate during exponential phase (up to day 6, Figure 4.21 or 4.49 A respectively) of 25-CHO-S/2.5C8/*cgr*TPRp-IRES-*cgr*Ttc36 was identical to 25-CHO-S/2.5C8/*cgr*TPRp, but at the subsequent stationary phase the cell density remained more stable and even slightly increased. The resulting change in integral viable cell density was 1.49-fold (≥ 80 % viability). Nevertheless, the overall growth as well as viability performance was not satisfying in comparison to sole overexpression of *cgr*Ttc36 (Figure 4.37). Hence, this combination and *cgr*TPRp alone (Chapter 4.3.1) were not further investigated.

The other combination with *cgr*Snord78 and *cgr*Ttc36 led to an even reduced growth as well as viability performance (Figure 4.49 B). Here, the intronic *cgr*Snord78 was situated upstream to *cgr*Ttc36 coding sequence (Table 3.21) and in comparison to the individual overexpressed factors (Figures 4.29 and 4.37), the combination was not desired. Somehow, both factors inhibited each other resulting in a 1.47-fold decrease in integral viable cell density. Therefore, the combination was not further elucidated, since at least an increase in maximal cell density was suggested.

5   Discussion

In following, the results as well as the state of the art are discussed and putative mechanisms behind the observed effects are elucidated. Due to the possible patent application, the factors TPRp, PPase and GTPase are discussed without mentioning the referred publications.

5.1   Increase of genomic variation by mutagenesis and the selection process

In following chapters, the performance of the applied mutagenesis and selection methods are discussed.

5.1.1   Mutagenesis and recovery of the 25-CHO-S cell line

The very first task was the generation of a clonal cell line suitable to grow at very high cell densities (up to $5 \cdot 10^7$ cells·ml^{-1}). Recent studies revealed the current achievements in maximal cell density ($2.07 \cdot 10^7$ cells·ml^{-1} for CHO-S cells and $1.70 \cdot 10^7$ cells·ml^{-1} for HEK293-S cells) (Beckmann *et al.* 2012; Liste-Calleja *et al.* 2013). These results were achieved by frequent passaging of the CHO-S cells leading to population selection (Beckmann *et al.* 2012) or by rational media optimisation avoiding nutrient limitations (Liste-Calleja *et al.* 2013). The study of Prentice *et al.* (2007) revealed the possibility to increase the viable as well as maximal cell density by stringently alternated fedbatch cultures. Due to rather low overall cell densities and the poor data due to the absence of replicates, the study showed some difficulties in the generation of very high cell density CHO-S cultures. Moreover, a similar stringent selection process on genetically unaltered CHO-S cells did not lead to the wished high cell density culture (personal communication), but to a reduced L-glutamine auxotrophy (Bort *et al.* 2010). Therefore, for elucidation of factors leading to increased cell densities the basal set of clonal variation within a cell line population was not sufficient such that the CHO-K1 derived suspension cell line needed to be genetically altered.

Again, the multiple ways of genetic variation could be performed such as chemically and physical mutations mentioned in chapter 2.2 or by epigenetic reprogramming (lost productivity by promoter methylation: Yang *et al.* 2010; Radhakrishnan *et al.* 2008). Mutagenesis by chemical mutagens were suggested to be the most efficient and easily applied method for the generation of new CHO-S phenotypes possessing at least one genetic modification (Chapter 2.2). Subsequently, the major questions to be answered were whether the inserted variations within the population of progenitor CHO-K1 cell line was sufficient to establish a high cell density growing clone as well as which selection procedure could be applied that is suitable to isolate such a clonal cell line.

The performed procedure of mutagenesis to increase the number of genotypic CHO-S variants was highly incidental. The used concentration of ethyl methanesulfonate (EMS) was previously optimised regarding a suitable cytotoxicity for the CHO-S cell lines. The mutagenic potency of EMS was suggested to be linear to the concentration within culture, since this parameter could not be measured. And recently, the added EMS concentration was proven to be suitable for insertion of deletions in CHO cells, rice plants and rice suspension cultures (Branda et al. 1999; Ribeiro et al. 2013; Chen et al. 2013b). Mutagenic potential could be measured by Comet assay or easily verified by micronuclei formation (Wagner et al. 2003). Here, the applied treatment of 400 µg·ml^{-1} EMS for 24 hours was sufficient to increase the ratio of micronuclei (Figure 4.6). In addition, the chromosomal disorders shown by micronuclei content were manifested during several passages suggesting a very efficient mutation process.

To further increase cell recovery after EMS-treatment and somehow preselect the cells, the freshly mutated cell population was treated by low temperature as well as by short-termed additions of ascorbic acid or conditioned medium from exponential growing 25-CHO-S cells (Figure 4.2). The latter supplement would possess cell specific growth factors and hormones providing anti-apoptotic stimuli and was therefore suggested to increase survival and recovery. In addition, treatment with ascorbic acid would intercept the reactive oxygen species (ROS) after EMS incubation avoiding necrotic or apoptotic cell death of putatively interesting mutants. Here, EMS was shown to increase ROS by glutathione depletion followed by ROS-damages and cell death (Tirmenstein et al. 2000). On the other hand, ascorbic acid was suitable to reduce intracellular ROS even after EMS-treatment (Cozzi et al. 1997; Kopnin et al. 2004). Furthermore, the decrease in culture temperature to 30-35 °C was reported to upregulate anti-apoptotic factors (Slikker et al. 2001; Fu et al. 2004). Taken together, all manipulations were applied to efficiently reduce necrosis as well as apoptosis and increase the mutated cell population. By applying a mean doubling time of 24 hours and a lack phase until day 2 (Figure 4.3), a total cell number of approximately 1'900 cells was estimated, which survived the mutagenesis and subsequent treatments inclusively cryopreservation.

To further reduce the cell number and increase the probability for isolation of putatively high cell density growing CHO-S clones, a novel selection procedure was performed.

5.1.2 Selection towards growth at high cell densities

The increase in cell density and growth rate of CHO-S-E400 in comparison to non-mutated 25-CHO-S cell line (Figures 4.4 and 4.5) was very promising regarding isolation of high cell density growing clones. Therefore, a preselection at culture conditions at higher cell densities (high osmolality and low L-glutamine) was performed (Figure 4.7). Here, selection at high osmolalities (450 – 550 mOsmol·kg^{-1}) led to more resistant CHO-DG44-S cell populations (Xiaoguang Liu 2010). Further, the absence of L-glutamine was simulated to enrich CHO(-K1)-S clones with ability of growing without L-glutamine, which was successfully shown before (Bort et al. 2010). The initial enrichment of ammonium during preselection process was avoided since the ammonium concentration was suggested to be increasingly produced during culture even without L-glutamine and by metabolisation of other amino acids (Bort et al. 2010). Additionally, ammonium has been shown to reduce glycosylation or its quality

(Zanghi et al. 1998; Borys et al. 1994). Therefore, clones with high ammonium resistance might be affected in glycosylation.

The preselection (Figure 4.7) was then performed with two serial batch cultures bearing 525 mOsmol·kg^{-1} (Osmolality of CM1035: approx. 300-350 mOsmol·kg^{-1}), absence in L-glutamine and until viabilities of approx. 50 %. The third round was performed as IS-fedbatch and with 600 mOsmol·kg^{-1} and 1 mM L-glutamine. In general, the osmolality was constantly increased during the culture period by putative metabolisation and evaporation (Table 3.4). Therefore, the effective osmolality was changed over time and possibly increase the selection pressure. Due to this suggestion and the rapid drop in viability during the second batch culture, the third preselection step was performed as fedbatch and slightly increased L-glutamine. Both changes should avoid further deprivation and slightly increase ammonium formation. Moreover, frequent feeding would forestall an osmolality far over 600 mOsmol·kg^{-1}, which was assumed to be toxic (IS CHO Feed-CD XP's osmolality: approx. 280-320 mOsmol·kg^{-1}).

The subsequently applied limited dilution to isolate the 156 CHO-S-E400$_{HyOsm}$ clones was performed with already feeded medium (Table 3.22) to increase the initial nutrient intake. Furthermore, evaporation effected an increase in osmolality and prolonged the respective selection pressure. Regarding the achieved clone number in comparison to the latter performed limited dilution experiments, the preselection was successful in respect to enrich resistant cell clones on high osmolality. Taken together, the isolated 156 CHO-S-E400$_{HyOsm}$ clones were estimated to approximately represent 8 % of the original cell number after recovery and would also include multiple clones due to overgrowth of poor growing clones. In addition, the preselection at high osmolalities might have reduced the clone number significantly.

The shaken batch selection rounds (Chapter 4.1.5) were applied simulating a long-term batch culture leading to cell lines with efficient nutrient consumption and therefore prolonged viable cell culture characteristics. This key selection was realised by visual viability and cell density estimation under comprehension of the aggregation status. This rough procedure was carried out to isolate clonal CHO-S-E400 cell lines with desired bioprocess characteristics only (low aggregation, high cell density, medium pH). Due to the gained experience with used cell type and medium, it was possible to visually distinguish the parameters. Even by thoroughly examination, potential cell lines with suitable features could be isolated by deeper view and determination of desired parameters (cell density, viability). Especially, fast growing cell clones with high peak cell densities but fast decline in viability could not be considered even they might surely possess interesting features. Nevertheless, out of 156 clones five cell lines survived and one CHO-S-E400 clone (1.104) was suitable for transcriptomic analysis. Moreover, this clone (CHO-S-E400/1.104) possessed identical aggregation as well as viability characteristics such as the clonal 25-CHO-S/2.5C8 cell line derived from 25-CHO-S. Solely, the maximal cell density and slightly the growth rate were largely increased (Figures 4.12, 4.14, 4.15 and 4.16). On the other hand, the ability for volumetric and specific productivity was significantly reduced, which was proven by transfection of hIgG$_A$ expression vector (Figure 4.13). Finally, this cell line pair was suggested to be perfectly matched and considered for external transcriptomic analysis.

5.2 Transcriptomic analysis and data verification

In chapter 2.4, different methods were presented for the identification of unknown cellular factors. All methods could not be performed locally and needed to be outsourced. Therefore, the first question revolved for specifying the most effective method. As shown in chapter 2.4.1, the overall protein content can be monitored and determined by proteomic analysis. This would exclude the regulatory RNA species such as e.g. microRNAs (miRNA), long non-coding RNAs (lncRNA) as well as small nucleolar RNAs (snoRNA). This species are of increasing interest and had to be included in our analysis. These points would also lead to the exclusion of transcriptomic microarray techniques due to limitations on known specific sequences and the absence of economical solutions as well as non-coding RNA platforms regarding applications with CHO sequences. Therefore, the use of sequential transcriptomic methods such as next-generation sequencing (NGS) as well as de-novo RNA sequencing methods were more suitable. Recently, a broad transcriptomic analysis by next-generation sequencing revealed and newly described several miRNA species in the Chinese hamster ovary cell lines CHO-K1 as well as CHO-DUXB11 for the first time (Hackl *et al.* 2011). At the moment of our differential transcriptomic analysis (early 2010) there was no open-source CHO sequence library available, hence the CHO sequences needed to be aligned to *R. norvegicus* or *M. musculus* sequence libraries. Due to the better basis on comparable published studies (Birzele *et al.* 2010; Yee *et al.* 2008), the HiSeq2000 NGS transcriptomic study was performed using the alignment to *M. musculus* genomic sequence library (ENSMUSG). Here, a high overall coverage of approx. 75 % could have been ensured, which could eventually lead to false results due to improper alignment. Overall, the RNA sequencing methods were suggested to be more robust and less error-prone due to mutation-related sequence alterations. The latter would have resulted to improper hybridisations in microarray-based transcriptomics.

The sample preparations were carefully performed to achieve proper results. The transcriptomic samples (Figure 4.16) were treated in an identical manner to ensure a maximal comparability. After next generation sequencing (NGS), the results were manually evaluated, whereby genes at $-2.0 \geq \log_2(\text{FoldChange}) \geq 2.0$ were considered only. Interestingly, the expression value/$\log_2$(FoldChange) plot (Figure 5.1) showed that low expression genes were rather upregulated than genes with higher initial expression. Overall, the expression of structural and bulk proteins was not significantly altered in CHO-S-E400/1.104. After further elimination of known modulators (Chapter 2.5), structural as well as bulk proteins, ribonucleoproteins and less quality reads, the number of ENSMUSG homologues was reduced to approx. 7 % of the original gene number. Subsequently, all genes were functionally characterised *in silico* and prioritised in respective to their p-value ($x \leq 0.001$), the possible significance for modulation growth at high cell densities as well as the documentation/publication status (Table 5.1). Thereupon, rarely published or functionally unknown targets were preferred to facilitate further potential patent applications.

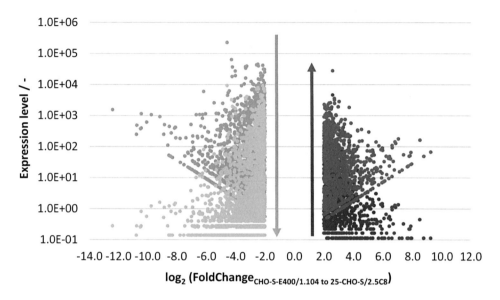

Figure 5.1: Overview of the differential transcriptome of CHO-S-E400/1.104 against 25-CHO-S/2.5C8. The expression level is plotted against the $\log_2(\text{FoldChange}_{1.104/2.5C8})$ showing the relationship and the frequency distribution. The darkened data points at each site represent the expression levels of the control cell line (25-CHO-S/2.5C8).

The thus selected factors (Tables 4.1 and 5.1) were subsequently verified by quantitative real-time PCR, regarding the partially observed massive variations in NGS differential expression levels within the biological replicates (e.g. for Gas5 or 9330182L06Rik in table 4.1). Indeed, significant expression discrepancies were observed (Figure 4.18) leading to the assumption that the transcriptome sequencing was not properly performed. Correlation of quantitative real-time PCR and transcriptomic analysis was observed for TPRp, 9330101J02Rik, PPase, GTPase as well as slightly for Hmga1 (Figure 4.18). The latter target was not significantly expressed ($1.0 \leq \log_2(\text{FoldChange}_{1.104/2.5C8}) \leq -1.0$) and therefore not further examined. Due to the significant overexpression of Gas5 and Ttc36, both were used for further studies, even the transcriptomic data was the total opposite. The targets without any changes in expression were Ankhd1, Btg3, Pdcd5, 9930012K11Rik, Filip1l and BC068281, whilst all targets showed high differential expression in transcriptomic data (Figure 4.18). In addition, targets with particularly strong media specific effects as were AI661453, miR-137, Cdkn1c and 9330182L06Rik, which were not suitable for proper expression profiling and therefore not capable for the further studies.

Table 5.1: Selected differentially transcribed factors in transcriptome analysis and their relevance in cancer research but no biotechnological relevance up to date.

TPRp †	- A tetratricopeptide-containing protein (anonymised by TPRp) - Overexpressed in several cancer tissues - Crucial for tumour proliferation - Predominant hypomethylation of genomic CpG islands results in overexpression
9330101J02Rik	- Functionally unrevealed gene - Homologous to *C. griseus* TPR repeat-containing protein C10orf93-like gene (LOC100759461)
miR-137	- Downregulated in several tumour spezies and mediates downregulation of a vast number of oncogenes (Althoff *et al.* 2013; Zhu *et al.* 2013; Luo *et al.* 2013a; Liang et al. 2013; Bi *et al.* 2013) - Downregulation is mediated by e.g. Hmga1 or promoter methylation (Liang et al. 2013; Vrba *et al.* 2013)
AI661453	- Functionally unknown theoretical *M. musculus* sequence - Homologous to R. norvegicus RGD1561662 gene (similar to AI661453 protein)
Ankhd1	- Ankyrin repeat and KH domain containing 1 - Overexpressed in some cancers and cancer cell lines and might be associated with p21 regulation (Traina *et al.* 2006; Dhyani *et al.* 2012) - Few publications available
Hmga1	- High mobility group AT-hook 1 - Overexpressed in several cancers and associated with cancer proliferation as well as metastasis ((Shah and Resar 2012); (Di Cello *et al.* 2013); (Shah *et al.* 2013); (Pegoraro *et al.* 2013) - Negatively regulates miR-137 (Liang et al. 2013)
BC068281	- Functionally unrevealed *M. musculus* sequence - Homologous to *C. griseus* WD repeat-containing protein C2orf44 LOC100760958
Ttc36 / HBP21	- Tetratricopeptide repeat domain 36 / HSP70 Binding Protein 21 - Binding to human HSP70 in breast cancer cells and proliferative vitreoretinopathy (Liu *et al.* 2008)
9330182L06Rik	- Functionally unknown theoretical *M. musculus* sequence - Homologous to *C. griseus* UPF0577 protein, KIAA1324-like homolog (LOC100760915) - Contains functionally uncharacterised intronic miRNA *mmu*-miR-879
Filip1l	- Filamin A interacting protein 1-like - Downregulation promotes either ovarian cancer proliferation and is associated with promoter methylation (Burton *et al.* 2011) or suppresses metastasis as well as cancer cell invasion (Kwon *et al.* 2013) - No correlation on expression status and global cancer phenotype

PPase †	- Reported to be downregulated in cancers and cancer cell lines - Connection to Src tyrosine kinase, focal adhesions and cell-cell as well as cell-matrix adhesion
Gas5	- Growth arrest-specific 5, long non-protein coding RNA (lncRNA) - Induces growth arrest in human and murine cell lines (Mourtada-Maarabouni and Williams 2013) - Frequent downregulation is associated with cancer promotion and cell proliferation (Liu *et al.* 2013c; Lu *et al.* 2013; Pickard *et al.* 2013; Qiao *et al.* 2013) - Contains several small nucleolar RNA species (Table 4.2; Shao *et al.* 2009) - Downregulated by miR-21 (Zhang *et al.* 2013), which is overexpressed at low temperatures in CHO-K1 (Gammell *et al.* 2007) and correlates with increase in viability
Pdcd5	- Programmed cell death 5 - siRNA knockdown results in apoptosis attenuation (Chen *et al.* 2006) - Downregulation is associated with cancer progression (Li *et al.* 2008; Xu *et al.* 2013) - Upregulation/Overexpression sensitises cancer cells and cell lines to apoptosis and inhibits proliferation (Fu *et al.* 2013; Xu *et al.* 2012a; Wang *et al.* 2013b)
Btg3	- BTG family, member 3 / B-cell translocation gene 3 - Negatively regulates the Src tyrosine kinase and the E2F1 transcription factor and is inhibited by p53 (Yu *et al.* 2008) - Reported to be suppressed in several cancer tissues and cell lines by promoter methylation and histone modification (Lv *et al.* 2013; Chen *et al.* 2013a; Deng *et al.* 2013; Majid *et al.* 2009)
9930012K11Rik	- Functionally unknown theoretical *M. musculus* sequence - Homologous to *R. norvegicus* RGD1308117 gene (similar to 9930012K11Rik protein)
Cdkn1c	- Cyclin-dependent kinase inhibitor 1C (p57, Kip2) - Cell cycle modulator, which decreases tumour cell growth when overexpressed (Giovannini *et al.* 2012), was documented to be frequently downregulated in several tumour species as well as mutated causing severe syndromes (Litvinov *et al.* 2012; Arboleda *et al.* 2012; Xu *et al.* 2012b) - Overexpression increased G1-phase fraction in HeLa and HEK293T cells (Hamajima *et al.* 2013) - Is destabilised/downregulated by Akt, whose overexpression was proven to increase CHO cell density (Zhao *et al.* 2013a; Hwang and Lee 2009)
GTPase †	- The downregulation of a homologous protein of identified GTPase leads to increased tumour cell migration and invasion ability. - Connection to Src tyrosine kinase, focal adhesions and cell-cell as well as cell-matrix adhesion

† TPRp, PPase and GTPase are anonymous replacements for known proteins and still under investigation

Taken together, due to the massive discrepancies in quantitative real-time PCR as well as transcriptomic analysis and the high quantity of quantitative real-time PCR expression data, the latter data were trusted more. Furthermore, by using samples of both fedbatch procedures (Figure 4.16) it was possible to show feed-dependent effects and further reduce the validation set, which led to more reliable results.

Since to the CHO genome database (Hammond *et al.* 2012a) was not available to date, the quantitative real-time PCR primers were designed based on *H. sapiens*, *R. norvegicus* and *M. musculus* sequence homology. This led to amplicons differing in size, but no disadvantage could be observed in amplification efficiency or else (data not shown). Additionally, the internal standard Actin B (ActB) had a consistent expression compared to the other housekeeping genes even during the growth phases (Figure 4.17). Nevertheless, using only one internal standard in quantitative real-time PCR was not the state of the art and error-prone, but was vindicated due to the consistent expression patterns of all tested housekeeping genes.

Due to the assumed poor quality of the transcriptomic data as well as the low throughput for quantitative real-time PCR verification, possible targets could not be gathered. Therefore, a putatively vast number of possible high cell density modulating target remained uncovered.

5.3 The tetratricopeptide repeat-containing proteins

The further subchapters are dealing with the examined tetratricopeptide repeat-containing proteins *cgr*Ttc36, *cgr*TPRp and *cgr*C10orf93-like/*cgr*Ttc40-like. The proteins are hardly (*cgr*Ttc36; Liu *et al.* 2008), slightly (*cgr*TPRp) or not documented (*cgr*Ttc40-like), therefore the discussion bases on assumptions and connections to other, better documented tetratricopeptide repeat-containing proteins. Thus, the goal is to dissect the differences in achieved cell densities (Figures 4.23, 4.24, 4.37, 4.39, 4.40 and 4.49) and to discuss common features such as the reduction of cell aggregation applying *cgr*Ttc36 and *cgr*C10orf93-like/*cgr*Ttc40-like (Figures 4.25, 4.41 and 4.45).

5.3.1 The effect upon the overexpression of *cgr*TPRp

The constitutive overexpression of the anonymised *C. griseus* tetratricopeptide repeat-containing protein *cgr*TPRp in non-producing 25-CHO-S/2.5C8 cell line resulted in no increase in maximal cell density (Figure 4.21) and abolished the growth beneficial effect of *cgr*Ttc36 when co-expressed (Figure 4.49). Moreover, the cell lines 25-CHO-S/2.5C8/*cgr*TPRp as well as 25-CHO-S/2.5C8/*cgr*TPRp-IRES-*cgr*Ttc36 showed decreased growth rates, but enhanced viability as well as viable cell culture characteristics at stationary phase compared to mock control cell line (Figures 4.21 and 4.49). The latter effect was even more enhanced during *cgr*TPRp and *cgr*Ttc36 co-expression, which was modulated rather by the viability increasing effect of *cgr*Ttc36 (Figure 4.37) than caused by *cgr*TPRp. Furthermore, the viability characteristics of mock cell line 25-CHO-S/2.5C8/CV121lin in these experiments showed an unusually early drop. Subsequent experiments (Figure 4.37, second fedbatch) were comparable with the observed viability and cell density stabilisation. Therefore, the constitutive overexpression of *cgr*TPRp was assumed to massively impair *cgr*Hsp70-binding proteins such as *cgr*Ttc36 and other proteins leading to the observed growth perturbation (Chapter 5.3.2.2).

In conclusion, due to the poor characteristics of *cgr*TPRp overexpression in 25-CHO-S/2.5C8 and the focus on more beneficial *cgr*Ttc36 the target was not further investigated concerning its influence on productivity. Finally, *cgr*TPRp could have positive impact on productivity at the end, but taken all observations together the volumetric productivity might be the same or even reduced once overexpressed in K20-3. Nonetheless, this assumption has to be proven further. The overexpression of *cgr*TPRp could at least help to elucidate the putative mechanism behind *cgr*Ttc36 (Chapter 5.3.2.2).

5.3.2 The putative function of *cgr*Ttc36

In chapter 4.3.4, the growth, cell density as well as viability enhancing effect of *cgr*Ttc36 in CHO-K1 derived clonal cell lines 25-CHO-S/2.5C8 and K20-3 were described. Here, the overexpression of *cgr*Ttc36 in non-producing 25-CHO-S/2.5C8 cell line resulted in a 1.78-fold to 1.88-fold increase in maximal cell density (Figure 4.37). This effect was slightly lower in hIgG-producing K20-3 cell line (Figures 4.39 and 4.40). In addition, the cell specific productivity was not perturbed nor increased by *cgr*Ttc36 overexpression in K20-3 (Chapter 4.3.4.3), but the inhibition of cell-cell aggregation was correlated to increased *cgr*Ttc36 mRNA content (Figures 4.45 and 4.46). Taken together, *cgr*Ttc36 overexpression could improve all production parameters (cell growth, viability, maximal cell density, integral viable cell density, culture duration and aggregation behaviour) but the cell specific productivity.

The sole documented function of human TTC36 (or HBP21) was the binding to heat-shock protein 70 (HSP70) yet (Liu *et al.* 2008). Consequential, the three documented tetratricopeptide repeat (TPR) domains (Figure 4.36) were crucial for binding of HSP70. In addition, *hsa*TTC36 extensively expressed in all examined breast cancer and slightly in normal tissues. Interestingly, *hsa*TTC36/HBP21 was tissue specifically truncated at the N- as well as C-terminus to HBP21s *in vivo*. The experimentally truncated HBP21s (compromising the three TPR domains, amino acids 51-156, only) possessed the highest binding affinity to HSP70 (Liu *et al.* 2008). Besides the binding of HSP70, the N- as well as C-termini of TTC36/Ttc36 might play a crucial role in regulation and signal transduction. Especially the C-terminal sequence downstream the three TRP domains revealed the highest homology within the examined species (LNPYAALCNRML) (Liu *et al.* 2008), which was predicted to be incorporated within a α-helix (Figure 5.2).

M	G	T	P	N	D	Q	A	V	L	Q	A	I	F	N	P	N	T	P	F	G
1	2	3	4	5	6	7	8	9	10	11	12	13	14	15	16	17	18	19	20	21

D	V	I	D	L	D	L	E	E	A	K	K	E	D	E	D	G	V	F	P	Q
22	23	24	25	26	27	28	29	30	31	32	33	34	35	36	37	38	39	40	41	42

TPRα

E	Q	L	E	Q	S	K	A	L	E	L	Q	G	V	R	A	A	E	A	G	D
43	44	45	46	47	48	49	50	51	52	53	54	55	56	57	58	59	60	61	62	63

TPRα   *TPRβ*

L	H	T	A	L	E	K	F	G	Q	A	I	C	L	L	P	E	R	A	S	A
64	65	66	67	68	69	70	71	72	73	74	75	76	77	78	79	80	81	82	83	84

TPRβ

Y	N	N	R	A	Q	A	R	R	L	Q	G	D	V	A	G	A	L	E	D	L
85	86	87	88	89	90	91	92	93	94	95	96	97	98	99	100	101	102	103	104	105

TPRβ   *TPRγ*

E	R	A	V	T	L	S	G	G	Q	G	R	A	A	R	Q	S	F	V	Q	R
106	107	108	109	110	111	112	113	114	115	116	117	118	119	120	121	122	123	124	125	126

TPRγ

G	L	L	A	R	L	Q	G	R	D	D	D	A	R	R	D	F	E	Q	A	A
127	128	129	130	131	132	133	134	135	136	137	138	139	140	141	142	143	144	145	146	147

TPRγ

R	L	G	S	P	F	A	R	R	Q	L	V	L	L	N	P	Y	A	A	L	C
148	149	150	151	152	153	154	155	156	157	158	159	160	161	162	163	164	165	166	167	168

N	R	M	L	A	D	M	M	G	Q	L	R	A	P	S	N	G	R
169	170	171	172	173	174	175	176	177	178	179	180	181	182	183	184	185	186

Figure 5.2: Secondary structure prediction of *cgr*Ttc36. Following features were predicted by NetSurfP (Petersen *et al.* 2009) as well as NetTurnP (Petersen *et al.* 2010) and highlighted as following: α-helices (red bars), β-turns (faint red bars) as well as coil structures (grey bars). β-Strands were not predicted. Bars were manually generated using the relative probability values (> 0.500) of raw data output. Exposed and buried residues are shown in black or grey letters respectively. Tetratricopeptide repeat (TPR) domains are shown in respective to (Liu *et al.* 2008) and figure 4.36.

Furthermore, it could not be deduced that the truncated HBP21s/TTC36s was predominantly formed in cancer cells. Thus, HBP21s/TTC36s rather possessed an irregular formation tendency suggesting a probable inhibitory effect to its parent molecule by reverse interception. The successive increase of the truncated TTC36/Ttc36 through parallel reduction in parent proteins would explain the rapid drop in cell density in *cgr*Ttc36 overexpressing CHO-K1 suspension cell derivatives (Chapters 4.3.4.2 and 4.3.4.4). Here, the cell densities dropped whilst the determined *cgr*Ttc36-mRNA ratios remained high in comparison to control cell lines. Interestingly, the increase in viability characteristics was virtually unchanged (data not shown) suggesting a rather post-translational or -transcriptional than epigenetic *cgr*Ttc36 regulation, whereat the latter was suggested to be possible due to the high CpG island content (Figure 4.38). Therefore, the putative secondary as well as tertiary structure, cellular localisation and modifications of *cgr*Ttc36 are further discussed to elucidate the possible mechanism in inhibition (Chapter 5.3.2.1). In chapter 5.3.4, the putative mechanisms of cell aggregation reduction by *cgr*Ttc36 as well as *cgr*C10orf93-like/*cgr*Ttc40-like are discussed. Furthermore, the discussion regarding the binding to Hsp70 and the resulting growth as well as viability increasing effect is performed (Chapter 5.3.2.2).

5.3.2.1   Prediction of secondary as well as tertiary structure, modifications and cellular localisation of *cgr*Ttc36

The putative protein structure of *cgr*Ttc36 was predicted using NetSurfP (Petersen *et al.* 2009) as well as NetTurnP (Petersen *et al.* 2010) leading to a predominant α-helices-bearing structure with intermediated β-turns (Figure 5.2). On the other hand, *cgr*Ttc36 showed very low probability for β-strand formation. The structure prediction perfectly matched with the tetratricopeptide repeat (TPR) domains formerly published of Liu *et al.* (2008). Here, each TPR domain compromised of two α-helices with a coil between and a β-turn at the C-terminus of each TPR domain (Figure 5.2). This secondary structure prediction was substantiated with x-ray or NMR data from homologous TPR-containing proteins predicted using PSI-Nature Structural Biology Knowledgebase (Gabanyi *et al.* 2011), in which the α-helices of each TPR domain were aligned in parallel and bridged by a coil structure (Figure 5.3). In addition, the surface accessibility prediction revealed the residues rather capable for interaction with binding partners as well as for post-translational modification. Furthermore, the differences between *cgr*Ttc36 and *hsa*TTC36 regarding their predictively exposed residues and secondary structures revealed large discrepancies between the N-terminal sequences (Figure 5.4). Here, the isoelectric points of both N-termini did not differ (*cgr*: 3.74; *hsa*: 3.67) even regarding the exposed residues only (*cgr*: 3.67; *hsa*: 3.60) and were shown to be highly acidic in comparison to the residual part (*cgr*: 10.30; *hsa*: 10.87) (Table 5.2). Furthermore, the TPRα and TPRβ domain were shown to be more acidic than the TPRγ domain as well as the C-terminus regarding both the total and exposed residues. The latter two basic domains or protein parts were suggested to putatively bind acidic protein sequences or even DNA. Nevertheless, the overall predicted secondary structures were shown to differ slightly only and were highly constant within the TPR domains and C-termini.

Taken together, the N-termini of both species revealed high variability beyond the highly homologous and structurally similar TPR domains as well as C-termini. The TPR domains were shown to interact with Hsp70, whereat the N-Termini simultaneously highly differ and the C-termini remained virtually unchanged during evolution from mollusc to mammals (Liu *et al.* 2008).   Thus, a main role of TPR domains and C-terminus in protein action as well as localisation could be suggested.

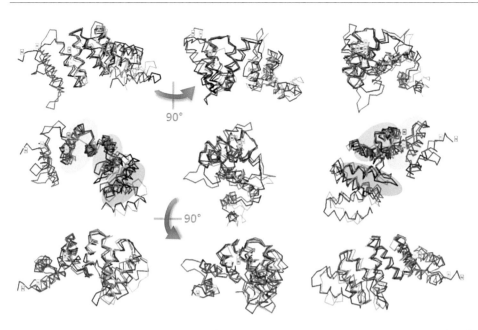

Figure 5.3: Tertiary structure prediction of monomeric *cgr*Ttc36 using PSI-Nature Structural Biology Knowledgebase (Gabanyi *et al.* 2011). Structures of four homologous proteins (PDB ID, sequence identity, position: 2y4tB, 30%, 51 – 151; 1w3bA, 22%, 6 – 183; 2vsnA, 21 %, 64 – 180; 3pe3B, 20%, 2 – 168) were overlaid resulting a tide structure prediction of *M. musculus* Ttc36 (*mmu*Ttc36, UniProt ac. Q8VBW8). The structures represent different views after angle alternation (90° and 180 ° in each direction respectively). The three tetratricopeptide repeat domains (TPRα – TPRγ) are showed in shaded red colours.

	12	13	14	15	16	17	18	19	20	21	22	23	24	25	26	27	28		29	30	31	32	33	34	35		
*cgr*Ttc36	A	I	F	N	P	N	T	P	F	G	D	V	I	D	L	D	L	-	E	E	A	K	K	E	D	-	-
*hsa*TTC36	A	I	F	N	P	D	T	P	F	G	D	I	V	G	L	D	L	G	E	E	A	E	K	E	E	R	E
	12	13	14	15	16	17	18	19	20	21	22	23	24	25	26	27	28	29	30	31	32	33	34	35	36	37	38

Figure 5.4: Differences in surface accessibility in *cgr*Ttc36 and *hsa*TTC36. Secondary structures and surface accessibilities were predicted by NetSurfP (Petersen *et al.* 2009) as well as NetTurnP (Petersen *et al.* 2010) and highlighted as following: α-helices (red bars), β-turns (faint red bars) as well as coil structures (grey bars). Exposed amino acid residues are shown in bold and buried residues in grey.

On the other hand, the N-terminal sequences might possess high relevance in regulation of the Ttc36/TTC36 proteins. For elucidation of the regulation sites, putative kinase-based phosphorylation sites, the serine/threonine modification with O-linked beta-N-acetylglucosamine (O-GlcNAc) as well as lysine modifications such as SUMOylation needed to be predicted. Additionally, the cellular localisation probability of *cgr*Ttc36 and *hsa*TTC36 was further elucidated.

Table 5.2: Calculated isoelectric points (pI) in cgrTtc36 and its domains. Sequences were taken from figure 5.2.

	All residues		Exposed residues	
	Sequence	pI	Sequence	pI
N-term	1 – 47	3.74	MGTPN-QA-Q-PN-GD-D-DLE-AK-ED-GV-QEQLE-	3.67
TPRα	48 – 81	5.88	-K---L------EAGD-HT--E--GQ--C-LPE-	4.42
TPRβ	82 – 115	6.25	-S---------RL-GD-AG--E--ER--T-SGG-	4.70
TPRγ	120 – 153	11.22	-Q---------RLQGR-DD--R--EQ--R-GSP-	10.00
C-term	154 – 186	11.41	-R----------A--NR--AD--GQ-RAPSNGR	12.02

The intracellular modification of serine and threonine residues with O-GlcNAc was shown to modulate cellular processes such as nuclear translocation, signal transduction, stress response and apoptosis (Ozcan *et al.* 2010; Zachara *et al.* 2011; Ma *et al.* 2013). Furthermore, O-GlcNAcylation was also shown to compete with phosphorylation sites, which might lead to the opposite effects (Tan *et al.* 2013; Ma *et al.* 2013). The prediction for a putative O-GlcNAcylation of cgrTtc36 and hsaTTC36 by YinOYang 1.2 (Gupta and Brunak 2002) revealed three significant serine or threonine residues modified in cgrTtc36 (T110, S122 and S183) and two in hsaTTC36 (S125 and S188). Nevertheless, either O-GlcNAcylation showed no significance to documented consensus sequence P-P-V-[ST]-T-A (Jochmann *et al.* 2013), suggesting a rather poor ability for O-GlcNAcylation.

The putative phosphorylation sites in cgrTtc36 and hsaTTC36 under consideration of their predicted structures revealed high probabilities for cAMP dependent protein kinase (PKA) mediated phosphorylation at serine residues 83/183 or 86/188, respectively (Table 5.3). Here, the PKA phosphorylation probability predicted by NetPhosK 1.0 (Blom *et al.* 2004) and NetPhos 2.0 (Blom *et al.* 1999) was even higher in hsaTTC36 than in cgrTtc36. Moreover, serine 151 in cgrTtc36 and serine 154 in hsaTTC36 were predicted to possess access to several kinases (ribosomal protein S6 kinase, p38 map kinase and cyclin-dependent kinase 5). Other putative phosphorylation sites were predicted by NetPhos 2.0 without specific kinase consensus sequence. The serine residues 112 or 115 were predicted to be most likely phosphorylated next to serine 83 / 86. Interestingly, the threonine 18 was shown to be exposed in hsaTTC36 and highly prone to be phosphorylated by p38 map kinase (p38MAPK). To complete the species' differences, CHO-K1 derived cell lines were shown to not express EGFR (epidermal growth factor receptor) (Krug *et al.* 2002). Even the ability for tyrosine 85 phosphorylation is theoretically given; this residue is not modified in native CHO-K1 derived cell lines.

Table 5.3: Putative phosphorylation sites in *cgr*Ttc36 and *hsa*TTC36 predicted by NetPhosK 1.0 (Blom *et al.* 2004) and NetPhos 2.0 (Blom *et al.* 1999). Predicted phosphorylation sites exposed residues are shown in bold letters.

cgrTtc36				**hsaTTC36**			
Residue	Kinase	Score (NetPhosK)	Score (NetPhos)	Residue	Kinase	Score (NetPhosK)	Score (NetPhos)
T 3	-	-	0.55	**T 3**	-	-	0.55
T 18	p38MAPK	0.53	0.72	**T 18**	p38MAPK	0.56	0.84
T 18	GSK3	0.50	0.72	-	-	-	-
T 18	Cdk5	0.51	0.72	-	-	-	-
T 66	CKII	0.51	0.02	**S 68**	CKII	0.53	0.05
S 83	PKA	0.62	0.98	**S 86**	PKA	0.62	0.98
Y 85	EGFR	0.58	0.63	Y 88	EGFR	0.58	0.63
T 110	-	-	0.61	-	-	-	-
S 112	-	-	0.89	**S 115**	-	-	0.98
S 122	PKC	0.74	0.38	S 125	PKC	0.74	0.38
S 122	PKA	0.60	0.38	S 125	PKA	0.61	0.38
S 151	RSK	0.52	0.81	**S 154**	RSK	0.53	0.81
S 151	p38MAPK	0.51	0.81	-	-	-	-
S 151	Cdk5	0.52	0.81	**S 154**	Cdk5	0.57	0.81
S 183	PKA	0.65	0.70	**S 188**	PKA	0.75	0.99

p38MAPK: p38 map kinase; GSK3: glycogen synthase kinase 3; Cdk5: cyclin-dependent kinase 5; CKII: casein kinase 2; PKA; cAMP dependent protein kinase; EGFR: epidermal growth factor receptor; PKC: protein kinase C; RSK: ribosomal protein S6 kinase. The score represents the phosphorylation probability for each kinase (Very low probability: 0.0; Highest probability: 1.00; Threshold: 0.50).

For elucidation of the putative structural changes due to phosphorylation, the mentioned serine/threonine residues were changed to asparagine *in silico* to simulate phosphoserine or phosphothreonine. The physiologically negatively charged asparagine residue was chosen due to the lowest possible pK_a in proteins (Taylor 1986); even the phosphorylated residues possess much lower pK_a constants (Andrew *et al.* 2002). If the pK_a might be lower, the constant depends on the residues microenvironment (Jaquet *et al.* 1993). This approximation might not describe real properties after phosphorylation, but should be suitable to show slight effects on *cgr*Ttc36 structure. Subsequently, the structures were predicted by NetSurfP (Petersen *et al.* 2009) as well as NetTurnP (Petersen *et al.* 2010) and manually aligned (Figure 5.5).

Interestingly, separate phosphorylation of serine residues at positions 83 and 183 did not lead to conformational changes, but the combined phosphorylation by the predicted cAMP dependent protein kinase (PKA) abolished the β-turn between positions 15 and 18 (Figure 5.5). Phosphorylations at threonine 110 and serine 151 were predicted to exhibit the same structural change, whilst winding the no-TPR domain α-helix at position 27-29 and 27, respectively. Additionally, phosphorylation at serine 112 was predicted to maintain the β-turn but unwrapping the α-helix at position 27-29.

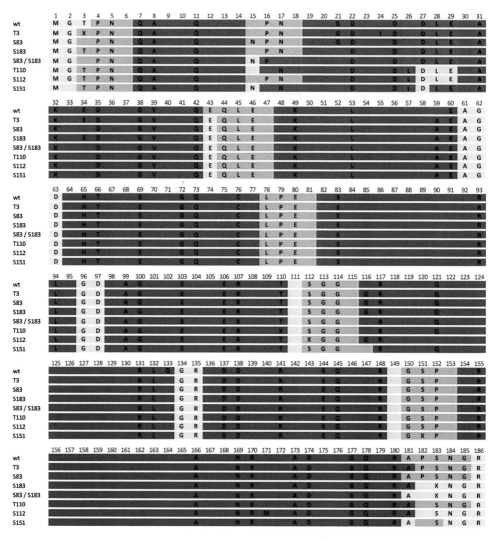

Figure 5.5: Putative structural changes in *cgr*Ttc36 after phosphorylation. At position X the residue was changed by aspartate (D) simulating the phosphorylation of serine/threonine. High impact N-terminal regions (15-18, 26-29 and 31-34) were highlighted in shaded grey. Following features were predicted by NetSurfP (Petersen *et al.* 2009) as well as NetTurnP (Petersen *et al.* 2010) and highlighted as following: α-helices (red bars), β-turns (faint red bars) as well as coil structures (grey bars). β-Strands were not predicted. Bars were manually generated using the relative probability values (> 0.500) of raw data output. Here, exposed residues are shown in black letters only.

Phosphorylation prediction at position 83, 83/183 and 112 revealed an altered C-terminal structure, namely the rather undefined coil structure formation. All these changes might cause changes in binding affinities as well as in modification accessibilities. Especially, the abolishment of β-turn in position 15 to 18 would lead to a massive change in tertiary structure

(Figure 5.3). This structural change would eventually cause loss or lowering in binding efficiency to Hsp70 or other proteins (Liu *et al.* 2008). Applying all predictions, the TPR domains (Figure 5.2) were maintained in structure and mainly in exposed residue content.

Next to putative phosphorylation sites, a likely Sumoylation site at position 32 was predicted by GPS SUMO 1.0 (Ren *et al.* 2009) and SUMOplot™ (http://www.abgent.com/sumoplot) (Table 5.4). The modification by SUMO (small ubiquitin-related modifier) is predominantly performed at lysine residues within type-I Ψ-K-X-E consensus sequences, whereas Ψ states for hydrophobic residues (Ren *et al.* 2009). Such a type-I consensus sequence was found in *cgr*Ttc36 at position 31-34, but not in *hsa*TTC36. Within the latter protein sequence the putatively SUMOylated lysine was changed to a glutamate residue (Figure 5.4), whilst shifting the SUMOylation site downstream to a rather improbable type-II non-consensus sequence (33-EKEE-36). The resulting Type-II non-consensus sequence was shown to be less effective in SUMOylation (Ren *et al.* 2009). Therefore and predicted by GPS SUMO 1.0, the human homologue might not be SUMOylated or at lower extent. Furthermore, the type-I consensus glutamate residue in position 36 was predicted to be buried within the α-helical structure or between two parallel α-helices (Figure 5.4).

Table 5.4: SUMOylation and SUMO binding sites in *cgr*Ttc36 and *hsa*TTC36 predicted by GPS SUMO 1.0 (Ren *et al.* 2009) and verified by SUMOplot™. The respective SUMOylation (lysine 32) and SUMO binding site are shown in red letters. The type-I consensus sequence (Ψ-K-X-E) is underlined. Thresholds were set to 5.86 (SUMOylation) and 70.1 (SUMO-binding).

	SUMOylation site		SUMO-binding	
cgrTtc36	DLDLEEAKKEDEDGV		SPFARRQLVLLNPYAALCN	
	Position	32	Position	158 – 162
	Score	17.75	Score	81.3
hsaTTC36	-		SPFARRQLVLLNPYAALCN	
	-	-	Position	161 – 165
	-	-	Score	81.3

A putative SUMOylation could lead to various mechanisms (Andreou and Tavernarakis 2009) including nuclear localisation as well as export (Santiago *et al.* 2013), modulation of ubiquitin-mediated degradation (Xing *et al.* 2012) and apoptosis (Li *et al.* 2013b; Liu *et al.* 2013a; Bettermann *et al.* 2012). On the other hand, SUMOylation was repotered to be regulated by e.g. phosphorylation (Qi *et al.* 2013; Lin *et al.* 2004). Interestingly, the accessibility of SUMOylation type-I consensus sequence in *cgr*Ttc36 (31-AKKE-34; Figure 5.2) changed after phosphorylation on serine residues 83, 112 as well as 151 and threonine residue 110 (Figure 5.5). Next to the structural change by abolishing the N-terminal β-turn, phosphorylation could lead to a change in SUMOylation efficiency. Due to the consensus glutamate internalisation, the SUMOylation efficiency was suggested to be rather reduced. In contrast, the ensuing amino acid residues were acidic or lipophilic, which was shown to be beneficial or even mandatory for efficient SUMOylation (Wang *et al.* 2013a; Yang *et al.* 2006). Therefore, the present predicted SUMOylation site in N-terminal *cgr*Ttc36 was suggested to be highly prone for SUMO modification, but could be regulated by phosphorylation.

Next to the modification by SUMOylating enzymes, *cgr*Ttc36 as well as *hsa*TTC36 were predicted to possess an identical SUMO binding site at positions 158 – 162 and 161 – 165 respectively (Table 5.4). Thus with high probability, the hamster Ttc36 would be capable to bind SUMOylated proteins as well as be modified by SUMO-specific E3 ligases, whilst human TTC36 would be able to bind SUMOylated proteins only. To elucidate the relevance of the present SUMO binding site, the sequences of several SUMO binding proteins were compared (Table 5.5).

Table 5.5: SUMO binding sites in *cgr*Ttc36 and other CHO-derived protein sequences predicted by GPS SUMO 1.0 (Ren *et al.* 2009). Basic (red letters), acidic (light red letters) and lipophilic (black letters) amino acid residues are highlighted. Hydrophilic as well as neutral amino acid residues are shown in grey letters. The height of consensus signs and legend letters represents the probability of occurrence. Score thresholds were set to 70.1 (high relevance).

Protein	Accession	Position	Sequence	Score
*cgr*Ttc36	experimental	158 – 162	SPFARRQ LVLLN PYAALCN	81.3
*cgr*Hsp70	AAA36991.1	334 – 338	DKSQIHD IVLVG GSTRIPK	87.2
		391 – 395	KSENVQD LLLLD VTPLSLG	72.6
*cgr*PIAS1	ERE76890.1	465 – 469	NKNKKVE VIDLT IDSSSDE	77.2
*cgr*PIAS3	ERE90883.1	369 – 373	KKAPYES LIIDG LFMEILN	72.9
		438 – 442	ENKKRVE VIDLT IESSSDE	76.3
*cgr*P53	CAA70109.1	252 – 256	GMNRRPI LTIIT LEDPSGN	70.7
*cgr*RanBP2	ERE87273.1	2357 – 2361	DPPSDTD VLIVY ELTPTPE	86.5
Consensus	+ (K/R/H) − (D/E) ○ (hydrophilic/neutral) ‡ (lipophilic/long-chained)		+−○‡ ‡‡‡‡○ ○‡−+	

Here, the core consensus sequence preferentially possessed four serial long-chained lipophilic amino acid residues (I/L/V) with a subsequent hydrophilic or neutral amino acid. The long-chained lipophilic amino acids were shown to be unequally crucial for a high SUMO-binding, which was shown recently (Song 2005) and by present alignment of CHO-derived protein sequences. Here, the second and forth (lipophilic) as well as fifth (neutral hydrophilic) amino acid within the core consensus revealed the highest relevance and alterations caused a massive loss in SUMO-binding efficiency. In addition, high content in hydrophilic residues within the near N-terminal and C-terminal sequences revealed higher scores in GPS SUMO 1.0 prediction. Interestingly, positively charged residues were rather found within N-terminal sequences and increased the scores even in altered consensus sequences, whilst C-terminal sequences predominantly possessed of neutral hydrophilic residues.

In conclusion, the predicted SUMO-binding score as well as the comparison with other hamster protein sequences of known or predicted SUMO-binding proteins (Wu and Chiang 2009; Werner *et al.* 2012; Brunet Simioni *et al.* 2009) revealed a high probability in binding SUMOylated proteins (Table 5.5). In contrast, the predicted SUMO-binding consensus sequence in *cgr*Ttc36 was shown to be buried within the C-terminal α-helix even after phosphorylation and therefore possibly not accessible (Figures 5.2 and 5.5). The SUMOylation and other possible as well as unrevealed modifications of *cgr*Ttc36 could result in structural changes such as increase in accessibility of the predicted SUMO-binding region. Thus, the role of SUMO modification as well as binding need to be further elucidated, but could also have impact on intracellular localisation. Consequentially, possible nuclear location as well as export sequences were described. Concerning the intracellular translocation of *cgr*Ttc36 and *hsa*TTC36, a possible nuclear export signal sequence was predicted by NetNES 1.1 (La Cour *et al.* 2004) as well as NESsential (Fu *et al.* 2011) from both species (Table 5.6).

Table 5.6: Predicted nuclear exports signals (NES) in *cgr*Ttc36 and *hsa*TTC36 by NetNES 1.1 (La Cour *et al.* 2004) and NESsential (Fu *et al.* 2011). The consensus residues are shown in red letters.

	NetNES 1.1		NESsential	
	Sequence	Score	Sequence	Probability
*cgr*Ttc36	102-LEDLERAVTL-111	0.50 – 0.54	At protein level	0.74
			20-FGDVIDL-26	0.31
			23-VIDLDL-28	0.48
			71-FGQAICL-77	0.29
			105-LERAVTL-111	0.99
*hsa*TTC36	105-LEDLERAVEL-114	0.35 – 0.41	At protein level	0.73
			20-FGDIVGL-26	0.19
			23-IVGLDL-28	0.56
			74-FGQAICL-80	0.30
			108-LERAVEL-114	0.94

The score regarding NetNES 1.1 prediction (for each residue) represents the quality of NES and the probability of cytosolic localisation (Very low probability: 0.0; Highest probability: 2.00; Default threshold: 0.50). The probabilities calculated by NESsential indicate the likelihood of a functional NES or the occurrence of a NES (In the range from 0 to 1).

This predicted data revealed high probabilities for a cytoplasmic localisation of *cgr*Ttc36 as well as *hsa*TTC36. By NEtNES 1.1 prediction, *cgr*Ttc36 showed a higher probability for the cytoplasmic localisation than *hsa*TTC36, whereas similar probabilities were found by NESsential prediction. Nevertheless, in both Ttc36/TTC36 species the nuclear export signals are buried within their secondary structure (Figure 5.2), whilst these were even more inaccessible in *hsa*TTC36 (Figure 5.4). Therefore, depending on reliability of the secondary structure prediction as well as the structural changes after possible posttranslational modifications and protein interactions, these buried export signals would become more accessible. At least phosphorylation was shown to negligible influence the predicted nuclear export signals. The putative *cgr*Ttc36 nuclear export signals between position 23 and 28 (VIDLDL) were predicted to be increased in accessibility after phosphorylation at positions 3, 110, 112 or 151 (Figure 5.5). Multiple modifications by phosphorylation among these serine or threonine residues were predicted to not further influence the probable nuclear export signals (predictions not shown). In contrast, putative nuclear localisation sequences in *cgr*Ttc36 and *hsa*TTC36 could not be predicted using several tools (NucPred and cNLS Mapper) (Brameier *et al.* 2007; Kosugi *et al.* 2009). Therefore, *cgr*Ttc36 as well as *hsa*TTC36 were predicted to be most likely localised within cytoplasm.

Furthermore, putative binding abilities and modifications other than the above characteristics of *cgr*Ttc36 and *hsa*TTC36 were examined. Firstly, following features were not predicted or probable for *cgr*Ttc36 nor *hsa*TTC36: Modification by a C-terminal glycosylphosphatidylinositol (GPI) anchor (PredGPI) (Pierleoni *et al.* 2008) or by palmitoylation (CSS-Palm 4.0) (Ren *et al.* 2008) as well as the appearance of signal peptides (SignalP 4.1) (Petersen *et al.* 2011) or transmembranal sequences (TMHMM Server v. 2.0) (Krogh *et al.* 2001). On the other hand, putative chaperone binding sites could be predicted using LIMBO molecular chaperone (Hsp70-specific) binding site prediction (van Durme *et al.* 2009). Equally in *cgr*Ttc36 and *hsa*TTC36, the chaperone binding sites were situated within the N-terminal part of TPR1 domain (SKALELQ) and with higher relevance within the C-terminal α-helical no-TPR structure (RRQLVLLNPY) (Table 5.7 and figure 5.2).

Table 5.7: Predicted chaperone binding sites in *cgr*Ttc36 and *hsa*TTC36 by LIMBO (van Durme *et al.* 2009).

	Position	Sequence	Score
***cgr*Ttc36** (*hsa*TTC36)	48 – 54 (51 – 57)	SKALELQ	8.8
	155 – 161 (158 – 164)	RRQLVLL	11.6
	157 – 163 (160 – 166)	QLVLLNP	15.7
	158 – 164 (161 – 167)	LVLLNPY	14.1

To elucidate the Hsp70-binding relevance, the predicted sequences were aligned to the CHO-K1 STIP1 homology and U box-containing protein 1-like or synonymously C-terminal Hsp70-interacting protein (*cgr*Chip; XP_003501570.1) an E3 ubiquitin ligase member, which was reported to bind the C-terminal (M/I)EEVD peptide of Hsp70/HSP70 with N-terminal TPR domain (Matsumura *et al.* 2013). Furthermore, *cgr*Chip was reported to mediate the degradation of e.g. c-Myc by interaction with Hsp70 (Paul *et al.* 2012). In addition, *cgr*Chip was suggested to mediate the turnover of Hsp70 itself (Mao *et al.* 2013). Interestingly, the predicted chaperone binding sequences of *cgr*Ttc36 at position 48 – 54 and 155 – 161 were shown to possess high sequence coverage with *cgr*Chip (Figure 5.6).

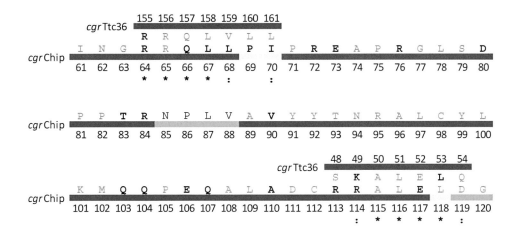

Figure 5.6: Sequence homology of predicted *cgr*Ttc36 chaperone binding sequences and *cgr*Chip (XP_003501570.1). Secondary structures and surface accessibilities were predicted by NetSurfP (Petersen *et al.* 2009) as well as NetTurnP (Petersen *et al.* 2010) and highlighted as following: α-helices (red bars), β-turns (faint red bars) as well as coil structures (grey bars). Exposed amino acid residues are shown in bold and buried residues in grey. Homologous residues are shown by asterisks (*) and residues with similar properties by colons (:).

By comparison the residue accessibilities of *cgr*Ttc36 and *cgr*Chip, the most probable chaperone binding sequence could be the N-terminal sequence of TPR1 domain (48-SKALELQ-54). Here, the other homologous chaperone binding sequence was predicted to be predominantly buried within the C-terminal α-helical structure subsequent to the TPR domains. This prediction was substantiated by the ulteriorly observed truncation of *hsa*TTC36 and the maintained binding capacity to HSP70 (Liu *et al.* 2008).

In summary, *cgr*Ttc36 was predicted to possess the almost identical structure as *hsa*TTC36 with the putatively same binding affinities to Hsp70. Phosphorylation as well as SUMOylation were outlined to be the major post-translational modifications with impact on structure and putatively the activity too. The tetratricopeptide repeat-containing protein *cgr*Ttc36 was predicted to be a soluble cytosolic molecule with a conserved tertiary structure (Figure 5.3) and to be capable of binding SUMOylated proteins. Within the following subchapters the possible influence of *cgr*Ttc36 (Chapters 5.3.2.2 and 5.3.2.3) as well as the putative mechanism behind the sudden growth inhibition (Chapter 5.3.2.4) are discussed.

5.3.2.2   The interaction of *cgr*Ttc36 with Hsp70 and the impact thereof

As mentioned above, *cgr*Ttc36 was shown to possess a strong structural relationship with *hsa*TTC36 (Chapter 5.3.2.1). Therefore and due to the predictions above (Chapter 5.3.2.1), it was assumed that *cgr*Ttc36 would also bind Hsp70 as reported similarly to *hsa*TTC36 (Liu *et* *al.* 2008). This suggestion was substantiated by previous studies on heat shock protein overexpression in interferon-γ producing CHO cell lines (Lee *et al.* 2009). Here, overexpression of hamster Hsp70 (*cgr*Hsp70), the possible binding partner of *cgr*Ttc36 was shown to increase the maximal viable cell density (1.80-fold) as well as the interferon-γ titer at viabilities above 80 % (2.67-fold) and to prolong the viable cell culture about two days. The grade of described cell density increase and viability prolongation was similar to the observed values for *cgr*Ttc36 overexpression in 25-CHO-S/2.5C8 (Figure 4.37). Since the reported data (Lee *et al.* 2009) was generated by using clonal interferon-γ producing CHO cell lines with putatively the highest *cgr*Hsp70 expression or best overall performance, the data was not fully comparable. Regarding the clonal variations in *cgr*Ttc36 overexpressing K20-3 cell clones (Figures 4.42 to 4.44), the same would have been observed in this previous publication and therefore it was suggested that the data was generated with the clone bearing the best characteristics. Nevertheless, it surely could not be denied that both data sets were surprisingly akin. In addition, the reported and performed data sets (Lee *et al.* 2009; Figures 4.37 to 4.44) revealed the independence of *cgr*Ttc36 or *cgr*Hsp70 action from cell specific productivity. Therefore, the major action of *cgr*Ttc36 was assumed to be the stabilisation of *cgr*Hsp70 leading to an improvement in cell density and viability without direct influence on productivity. Since the *cgr*Hsp70 expression status was not analysed in this thesis, it could not be monitored if the cellular differences between 25-CHO-S/2.5C8 as well as K20-3 cell line and even clonal differences were due to the *cgr*Hsp70 content, which was assumed. This suggestion could be substantiated by performed 25-CHO-S/2.5C8/*cgr*Ttc36 cultures in absence of glutamine, whereat the increase in cell density was abolished and levelled to control (data not shown). Glutamine was reported to increase the Hsp70 level in murine embryonic fibroblasts (Hamiel *et al.* 2009). Hence, the increase in growth, viability as well as cell density mediated by the overexpression of *cgr*Ttc36 was assumed to be directly linked to *cgr*Hsp70 expression status.

The hypothesis of direct linkage of *cgr*Ttc36 with *cgr*Hsp70 was substantiated by the performed co-expression of *cgr*TPRp with *cgr*Ttc36 (Figure 4.49 A). The observed abolishment of *cgr*Ttc36-driven high-density growth could theoretically be predicted on the putative impairment of the *cgr*Ttc36-*cgr*Hsp70 interaction by *cgr*TPRp directly. Hence, the interception of *cgr*Hsp70 by *cgr*TPRp could further avoid beneficial interaction with other proteins. Therefore, the excessed appearance of *cgr*TPRp by overexpression could be capable to compete the interaction of *cgr*Ttc36 and other proteins to *cgr*Hsp70 resulting in the observed growth inhibition. Interestingly, the chaperone-binding sites of *cgr*Ttc36 and *cgr*Chip are very similar, whilst *cgr*TPRp partly differs from both sequences (Figure 5.6; Table 5.8). Hence, the capability of the possible competition of *cgr*Chip by *cgr*Ttc36 stabilising *cgr*Hsp70 is obvious. Nevertheless, since it was experimentally shown that *cgr*TPRp is able to compete this effect, either the expression level of *cgr*TPRp is higher in 25-CHO-S/2.5C8/*cgr*TPRp-IRES-*cgr*Ttc36 or *cgr*TPRp possesses a higher specific for *cgr*Hsp70.

Driven by vector construct CV147, which encodes for *cgr*TPRp as well as *cgr*Ttc36 (Table 3.21), *cgr*TPRp is possibly translated at higher rates than *cgr*Ttc36. Shown for BHK21 cell lines, genes located downstream of the internal ribosomal entry site (IRES$_{FMDV}$) were expressed with an approx. 4-fold lower translation efficiency (Wong *et al.* 2002; Marschalek *et al.* 2009). Suggesting a similar transcription level for the *cgr*TPRp-IRES-*cgr*Ttc36 mRNA than for *cgr*Ttc36 mRNA alone (approx. 64-fold, Figure 4.46), both proteins would compete intrinsic chaperone-binding proteins in a similar manner, whereby this would not explain the total abolishment of the effect of *cgr*Ttc36 by simply 4-fold excess of *cgr*TPRp. Hence, the initial [R/K][R/K] pair followed by hydrophobic residues could mediate the binding to the C-terminal (M/I)EEVD as well as PTIEEVD peptides of Hsp70/HSP70, which was found to be essential for CHIP binding (Matsumura *et al.* 2013) (Table 5.8). The initial serine-lysine (SKALELQ) in *cgr*Ttc36 could be less efficient in binding of *cgr*Hsp70 than the lysine-lysine in *cgr*TPRp. Thus, the competition of *cgr*Ttc36 by *cgr*TPRp might be explained by the higher translation rate of *cgr*TPRp compared to *cgr*Ttc36 applying the CV147 vector construct and by the possible higher binding affinity of *cgr*TPRp to *cgr*Hsp70.

Table 5.8: Chaperone binding site alignment of *cgr*Ttc36 (experimental sequence), *cgr*Chip (XP_003501570.1) and *cgr*TPRp (anonymised). Identical as well as similar residues are highlighted in shades of red.

	Predicted chaperone binding sites	
cgrTtc36	48-SKALELQ-54	155-RRQLVLL-161
cgrChip	113-RRALELD-119	64-RRQLLPI-70
cgrTPRp	5-KKxxxxx-11	96-RKxxxxx-102

A similar mechanism was recently shown by OLA1 (Obg-like ATPase 1) overexpression, which was demonstrated to regulate the HSP70 turnover by interfering binding of HSP70 to CHIP (C-terminus of Hsp70-binding protein) (Mao *et al.* 2013). CHIP was reported to mediate the degradation of HSP70/Hsp70 by ubiquitination (Jiang *et al.* 2001). Thus, competition of CHIP/*cgr*Chip would increase the HSP70/Hsp70 content by its stabilisation, which was proven by OLA1 overexpression (Mao *et al.* 2013).

Taken together, *cgr*Ttc36 was predicted to be highly prone to bind and stabilise *cgr*Hsp70 simultaneously to OLA1 (Mao *et al.* 2013), which is illustrated by effect similarity of *cgr*Ttc36 as well as *cgr*Hsp70 overexpression in serum-free CHO-K1 suspension cultures (Lee *et al.* 2009) and the putative inhibition of this effects by *cgr*TPRp overexpression. Therefore, the growth-beneficial effects of *cgr*Ttc36 were suggested to be directly linked with the expression as well as activity of *cgr*Hsp70. Nevertheless, other signal transduction mechanisms and binding partners could be possible, especially regarding the observed maintenance of prolonged viability after freezing/thawing of *cgr*Ttc36 overexpressing 25-CHO-S/2.5C8 cells (data not shown, chapter 4.3.4.2) as well as during co-expression with *cgr*TPRp (Figure 4.49 A) and the aggregation inhibition (Figure 4.41). Subsequently, the role of the possible intrinsically generated fusion protein *cgr*KmtTtc36 and further inhibition mechanisms are discussed in chapters 5.3.2.3 and 5.3.2.4, respectively.

5.3.2.3 The mechanism behind *cgr*KmtTtc36 fusion protein formation and action

A very interesting appearance and a possible linkage to more persistent DNA methylation-independent epigenetic mechanisms was found applying different genomic and transcribed sequences available upstream *cgr*Ttc36. In recently sequenced *C. griseus* sequences only (strain 17A/GY chromosome 4 unlocalised genomic scaffold chr4_scaffold_52), a fusion protein (ERE74997.1; gi|537176041) compromised of histone-lysine N-methyltransferase MLL1 (gi|100760424) with downstream *cgr*Ttc36 (gi|100758106) was observed (further called *cgr*KmtTtc36). In *H. sapiens* (gi|4297), *M. musculus* (gi|214162) as well as *R. norvegicus* (gi|315606) the MLL1-homologous histone-lysine N-methyltransferase 2a (Kmt2a/KMT2A) (further called *cgr*Kmt2a/MLL1) was found to be in tight vicinity upstream of the respective Ttc36/TTC36 locus. In earlier CHO-K1 based sequencing data, the *cgr*Kmt2a/MLL1 (XM_003510519.1) as well as *cgr*Ttc36 (XM_003510512.1) were found to be transcribed separately. Therefore and due to the short genomic distance between *cgr*Kmt2a/MLL1 -TAA as well as *cgr*Ttc36-ATG (5130 bp, based on gi|344163747), an alternative splicing event was hypothesised. Therefore, the *cgr*KmtTtc36 mRNA (ERE74997.1; gi|537176041) was compared by alignment of CHO-K1 unplaced genomic scaffold2149 whole genome sequence (gi|344163747; gb|JH001318.1), the *cgr*Kmt2a/MLL1 coding sequence (XM_003510519.1; gi|354496910) and the documented sequence of *cgr*Ttc36 (XM_003510512.1; gi|354496896:2-562) (Figure 5.7).

Thereby, common features such as splice consensus sequences explained the mechanism of *cgr*KmtTtc36 formation. In addition, the stop codon of *cgr*Kmt2a/MLL1 (TAA) is embedded within the first 13 bp of the intronic 5115 bp sequence between *cgr*Kmt2a/MLL1 and *cgr*Ttc36. This stop codon could be removed by splicing, whilst the *cgr*Kmt2a/MLL1 as well as *cgr*Ttc36 coding sequences are fused. The intronic consensus sequences were shown to be efficient enough for splicing since the necessary intronic elements are present (5'-|GT and TAG|-3') even the consensus 3'-splice site YAG|G is exchanged to putatively less efficient TAG|T sequence (Figure 5.7) (Ast 2004; Rogers and Wall 1980). Promoter sequences within the intron upstream of the *cgr*Ttc36 coding sequence (gi|344163747; from 182324), were capable for abolishment by splicing as well and were predicted by Promoter 2.0 (Knudsen 1999) under consideration of scores higher than 1.00 only.

The subsequent promoter sequence alignment of homologous intronic sequences derived from genomic sequences of *C. griseus* (gi|344163747|gb|JH001318.1), *M. musculus* (gi|372099101:c44893059-44791653), *R. norvegicus* (gi|389675121:c47815895-47717569) and *H. sapiens* (gi|224589802:118293654-118411088) revealed three TATAAA sequences for *C. griseus* Kmt2a/MLL1-Ttc36 connecting intron (gi|344163747:180562-180567; 180939-180944; 180953-180958) compared to two widely separated TATA-boxes in *M. musculus* and each one in *R. norvegicus* and *H. sapiens*.

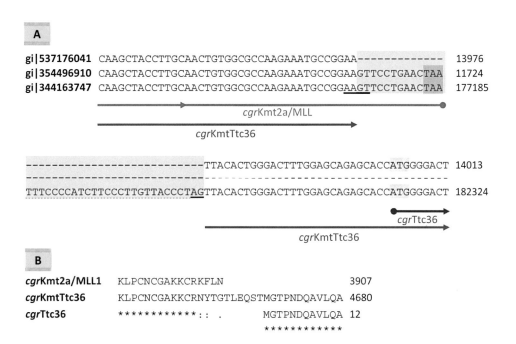

Figure 5.7: Sequence alignment of *cgr*KmtTtc36 mRNA (gi|537176041) with CHO-K1 unplaced genomic scaffold2149 whole genome sequence (gi|344163747) as well as *cgr*Kmt2a/MLL1 coding sequence (gi|354496910) (A) and the comparison of resulting change in protein sequence of *cgr*KmtTtc36 (B). The observed intronic sequence is shaded by light red. The arrows indicate the coding sequences of the respective protein species. Stop as well as start codons are shaded in grey. The intronic consensus sequences (5'-GAA|GTTCCT as well as TTCCCCATCTTCCCTTGTTACCCTAG|TTA-3' are highlighted as following: splice consensus sites; poly(C/T) tail).

Therefore, the CHO-K1 sequence revealed a higher content of putatively active TATA-boxes than in other species suggesting a more prominent expression of Ttc36 mRNA in *C. griseus* than in the other examined species. In addition, the promoter sequences necessary for constitutive expression (CCAAT and GC-rich boxes; Knudsen 1999) were found to be upstream of each of the three TATAAA sequences, suggesting that the expression of *cgr*Ttc36 might be higher than the expression of *cgr*KmtTtc36 (Figure 5.8).

Furthermore, putative heat-shock response elements (alternating purine and pyrimidine-rich sequences, e.g. GAANNTTC or TTCNNGAA; Amin *et al.* 1988; Anckar *et al.* 2006) were observed upstream the predicted TATA-boxes, assuming an induced or increased expression by heat shock. Another feature within the related *H. sapiens* intronic sequence only, was observed to be aryl hydrocarbon receptor-inducible xenobiotic/dioxine responsive elements (TNGCGTG; Dere *et al.* 2011) suggesting an autonomous mechanism in evolution of *H. sapiens* recruiting *hsa*TTC36 as an additional stress responsive as well as protective protein against xenobiotics.

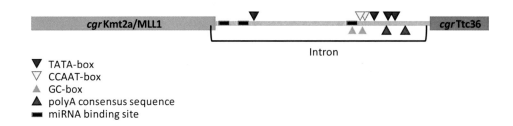

▼ TATA-box
▽ CCAAT-box
▲ GC-box
▲ polyA consensus sequence
━ miRNA binding site

Figure 5.8: Schematic layout of promoter sequences, polyA consensus sequences and miRNA binding sites observed within the internal *cgrKmt2a/MLL1-cgr*Ttc36 intronic sequence. Coding sequences of *cgrKmt2a/MLL1* and *cgr*Ttc36 are shown in red shaded blocks, whereas intronic sequence is flanked by the bracket.

The monomeric non-fusion protein *cgr*Kmt2a/MLL1 plays an important role in histone 3 lysine 4 trimethylation (H3K4me3), which is crucially mediated by MLL1-WDR5 (WD repeat domain 5) interaction and supported by RbBP5 (retinoblastoma binding protein 5) as well as Ash2L (ash2 (absent, small, or homeotic)-like) (Dou *et al.* 2006; Crawford and Hess 2006). H3K4me3 genomic regions are strongly associated with active promoters and strong enhancers (Pekowska *et al.* 2011). In this order, H3K4me3 residues were found within the CMV promoter region of stably transfected luciferase vectors (Okitsu *et al.* 2010). MLL1 possesses high specificity as well as affinity for methylated H3K4 residues, whereat the dissociation constant (K_D) was lower the higher grade in methylated occurred. Additionally, this binding to H3K4me3 is crucial for MLL/WDR5/RbBP5/AshL2 mediated transcription stimulation (Chang *et al.* 2010). This finding is coherent with the pattern in CMV promoters, whilst H3K4me3 content was 5- to 7-fold higher than H3K4me2 content suggesting a crucial role of H3K4me3 in transcription activation (Okitsu *et al.* 2010). In addition, these modifications were shown to be excluded from methylated DNA regions substantiating the function in actively transcribed regions (Okitsu and Hsieh 2007). The H3K4 bi- as well as trimethylation associated with active promoters was found to impair H3K9 methylation mediated by e.g. SUV39H1 (suppressor of variegation 3-9 homolog 1) as well as SETDB1 (SET domain, bifurcated 1), which predominantly led to gene silencing (Binda *et al.* 2010). The latter histone trimethylation (H3K9me3) subsequent to H3K9 deacetylation by HDAC1 (histone deacetylase 1) was shown to be crucial for proper chromosome condensation and gene silencing (Park *et al.* 2011). This was further reported to cause cyclin B1 reduction in oxidatively stressed HeLa human cervical adenocarcinoma cell line and could cause growth as well as G2 cell cycle arrest (Chuang *et al.* 2011). On the other hand, histone deacetylase inhibitors mediated the H3K9me3 demethylation via e.g. suppression of SUV39H1 leading to a reactivation of prior silenced promoters or genomic regions (Wu *et al.* 2008). Subsequently, histone acetylation after HDAC1 inhibition initiates H3K4 methylation as well as transcription activation therefore (Marinova *et al.* 2011).

In order of intronic sequence analysis on putative microRNA (miRNA) binding sites by DIANA-microT v3.0 (Maragkakis *et al.* 2009), the miRNAs *cgr*-miR-125b-5p, *cgr*-miR-429 and *cgr*-miR-146b-3p were predicted as regulators of *cgr*Kmt2a/MLL1, *cgr*Ttc36 and possibly *cgr*KmtTtc36 as well (Figure 5.8; Table 5.9). Due to the proximity of *cgr*-miR-125b-5p binding site to *cgr*Kmt2a/MLL1 stop codon, this matured miRNA species was suggested to negatively regulate *cgr*Kmt2a/MLL1 translation by mRNA destabilisation. In addition, the putative *cgr*-miR-429 binding site was found within the early 3'-UTR (untranslated region) of *cgr*Kmt2a/MLL1, which possesses a size of 3748 bp between stop codon and first polyA signal (Figure 5.8). For both miRNAs functionally similar targets were predicted (Table 5.9). Focusing on epigenetic modifiers, *cgr*-miR-125b-5p and *cgr*-miR-429 showed specificities against initiating (MLL5) as well as repressive H3 methyltransferases (SUV39H1 and SUV420H2) and their associated modulators (BCL11B/CTIP2) as well as modulators of histone deacetylase 1 (HDAC1) (SMEK1 and FBXW7). SUV39H1 was experimentally proven to be downregulated by miR-125b in mice and human (Villeneuve *et al.* 2010; Fan *et al.* 2013). Therefore, downregulation of *cgr*Suv39H1 by *cgr*-miR-125b-5p might be very likely due to this interspecies coincidence. The H4K20 specific methyltransferase SUV420H2 is involved in gene repression as well (Kapoor-Vazirani *et al.* 2011). Binding site prediction revealed *cgr*-miR-429 mediated BCL11B/CTIP2 posttranscriptional downregulation, which is associated with HDAC1, HDAC2 as well as SUV39H1 and mediates H3K9 deacetylation/methylation (Marban *et al.* 2007). In parallel, SMEK1 (SMEK homolog 1, suppressor of mek1) interacts with HDAC1, which leads to deacetylated histone residues in target regions (Lyu *et al.* 2011). The predicted miR-429 target FBXW7 (F-box and WD repeat domain containing 7, E3 ubiquitin protein ligase) was reported to stabilise histone deacetylation, thus loss in FBXW7 led to higher HDAC1/2 inhibition efficiency (Yokobori *et al.* 2014; He *et al.* 2013).

Next to putative posttranscriptional inhibition of *cgr*Kmt2a/MLL1 and predicted inhibitory binding of MLL5, *cgr*-miR-125b-5p and *cgr*-miR-429 thus negatively influence the repressive histone-specific epigenetic modifications as well. Therefore, the overall histone acetylation/methylation system might be changed after parallel transcription of both miRNAs. The nuclear translocation protein KPNA6 (karyopherin α6/Importin α7) was predicted to be a target of *cgr*-miR-125b-5p suggesting the partial and transient abolishment of nuclear transit. This nuclear transport protein facilitates the nuclear import of oxidative stress-induced Keap1 within the nucleus, whereat downregulation of KPNA6 resulted in nuclear import inhibition (Sun *et al.* 2011). The third predicted miRNA *cgr*-miR-146b-3p could target KLHL21 (kelch-like 21), which is required for mitosis progression and cytokinesis. Downregulation of KLHL21 could induce a growth arrest therefore (Maerki *et al.* 2009). The miRNA *cgr*-miR-146b-3p was predicted to bind within the promoter sequences of the *cgr*Kmt2a/MLL1-*cgr*Ttc36 intron (Figure 5.8). Thus, a posttranscriptional inhibition of *cgr*Kmt2a/MLL1 as well as transcription inhibition of *cgr*Ttc36 through binding on its genomic promoter GC-box sequence is assumed.

Table 5.9: Predicted as well as putative binding of *cgr*-miR-125b-5p (MIMAT0023743), *cgr*-miR-429 (MIMAT0023960) and *cgr*-miR-146b-3p (MIMAT0023776) to the connecting intron on *cgr*Kmt2a/MLL1-Ttc36 pre-mRNA. MicroRNA sequences were derived from miRBase Blast results (Griffiths-Jones *et al.* 2006; Griffiths-Jones *et al.* 2007). Pre-mRNA sequences derived from genomic sequence (gi|344163747|gb|JH001318.1). Binding and coverage are shown in red lines. Selected targets were predicted by DIANA-microT v3.0 (Maragkakis *et al.* 2009) based on human as well as murine microRNA homologue and selected for incorporation in histone modification processes.

	Sequence alignment		Selected miRNA targets
cgr-miR-125b-5p	22 AGUGUUCAAUCCCAGAGUCCCU 1 ||| || |||| |||| ||| 177253 UCAGCAGCUAGGAUCUCCAGGA 177274	Epigenetic markers	SMEK1 SUV39H1 SUV420H2
		Nuclear import	KPNA6
cgr-miR-429	22 UGCCGUAAUGGUCUGUCAUAAU 1 | ||| |||||||| ||| 178046 UGGCCAUAACCAGACAGAAUUC 178067	Epigenetic markers	BCL11B/CTIP2 FBXW7 MLL5/KMT2E
cgr-miR-146b-3p	22 UGGUCUUGACUCAGGGAUCCCG 1 |||||||| ||||| || 180283 GCCAGAACUUGGUCCCUGGGCT 180304	Mitosis	KLHL21

Taken together, *cgr*-miR-125b-5p and *cgr*-miR-429 are inducible miRNA species with common features and potential for inhibition of *cgr*Kmt2a/MLL1 translation and its transport within the nucleus as well. The spliced *cgr*KmtTtc36 mRNA would not be accessible for this regulatory mechanism, therefore the role of this fusion protein should now be elucidated.

Due to the structural findings above and splicing events, *cgr*Kmt2a/MLL1 is strongly suggested to be inactivated and downregulated by stress responses, whereat *cgr*KmtTt36 would not be influenced. Moreover, the promoter region of *cgr*Kmt2a/MLL1 (5'UTR) revealed a rather constitutive expression with two proximal TATA-boxes, two CCAAT-boxes and no special regulatory elements but two GC-boxes (Sp1 consensus binding sites; Segal *et al.* 1999). Expression MLL1 was shown to be induced by stress-responses such as toxins via transcription factor Sp1 binding on MLL1-promoter region (Ansari *et al.* 2009). Therefore, *cgr*Kmt2a/MLL1 is suggested to be constitutively expressed, upregulated upon toxin response and predominantly downregulated by its 3'UTR as well as by other stress responses. Interestingly, two putative NFκB1/RelA (p50/p65) binding sites were found on the donor splice site and within the intronic sequence either (Figure 5.9) (Wong *et al.* 2011). This means on the one hand that the expression of *cgr*Ttc36 is crucially controlled upon NF-κB pathway activation, in which RelA (p65) homo- or heterodimers with NFκB1 (p50/65) are phosphorylated and subsequent histone residues demethylated as well as acetylated (Yang and Chen 2011). Regarding the transcription in CHO-S-E400/1.104 depending on the culture duration (Figure 4.19), cgrTtc36 was downregulated in exponential phase, whilst upregulation occurred at early stationary phase. This finding could be substantiated by NFκB1/RelA mediated activation of cgrTtc36 transcription, since IκBα was downregulated 3.21-fold in CHO-S-E400/1.104 compared to control cell line (transcriptomic data not shown, p-value = $3.27 \cdot 10^{-7}$), which is a negative regulator of the NF-κB pathway transduction (Tang $et~al.$ 2013). In addition, the first NFκB1/RelA binding sites form the donor splice site of cgrKmt2a/MLL1-cgrTtc36 intron, which would possibly influence splicing efficiency upon NF-κB pathway activation. Underlining this suggestion, p50 homo-/heterodimers were found to interact with RNA but with lower efficiencies (Khan $et~al.$ 2012).

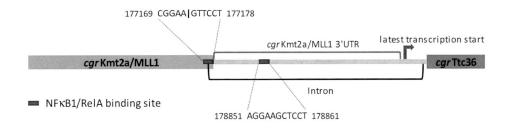

Figure 5.9: Schematic overview on cgrKmt2a/MLL1 and cgrTtc36 genomic elements including putative NFκB1/RelA binding sites. The respective NFκB1/RelA (p50/p65) binding sequences were predicted using recent consensus sequence optimisation (Wong $et~al.$ 2011) and are highlighted with their position with CHO-K1 unplaced genomic scaffold2149 whole genome sequence (gi|344163747). The first NFκB1/RelA binding site is additionally displayed with the predicted splice site.

Therefore, it is suggested that splicing would occur after RelA homo-/heterodimers dissociation and would be inhibited by NF-κB pathway activation, whereby the sequences possess higher probability for RelA homodimer binding (Wong $et~al.$ 2011). Furthermore, the cgrKmt2a/MLL1 posttranscriptional inhibitor cgr-miR-125b-5p is upregulated by oxidative stress responses suggesting a subsequent reduction of cgrKmt2a/MLL1 mRNA (Joo $et~al.$ 2013). In addition, UV-mediated activation of the NF-κB signalling lead to upregulation of miR-125b-5p in HEK293 and HaCaT cells (Tan $et~al.$ 2012). Furthermore, LPS treatment of human biliary epithelial H69 cells and subsequent NF-κB pathway onset temporarily increased miR-125b transcription leading to the suggestion that cgrKmt2a/MLL1 is strongly regulated by oxidative as well as inflammatory signals (Zhou $et~al.$ 2010). In contrast, MLL1-deficient mouse embryonic fibroblasts revealed a perturbed NF-κB pathway transduction after TNFα or LPS treatment by reduced H3K4 methylation status (Wang $et~al.$ 2012). This would implement a feedback mechanism of cgrKmt2a/MLL1 on NF-κB pathway assuming that both the epigenetic modulation as well as inflammatory NF-κB pathway would regulate each other avoiding either overregulation.

Taken together, *cgr*Ttc36 is assumed to be overexpressed upon heat-shock, Sp1 as well as NF-κB pathway activation, whilst a basal expression level would be maintained due to present promoter elements (Figure 5.8). Expression of *cgr*Ttc36 is stopped after UTR-intron splicing and fusion to *cgr*Kmt2a/MLL1. This mechanism is assumed to be directly inhibited by RelA homo-/heterodimer binding on donor splice site with subsequent increase of *cgr*Kmt2a/MLL1 3'UTR accessibility and downregulation by *cgr*-miR-125b-5p thereof (Figures 5.8 and 5.9). The overexpression of *cgr*-miR-125b-5p as well as *cgr*-miR-429 would additionally lead to a reduction in H3K4 methylation (MLL1/MLL5 inhibition), H3 acetylation (indirect inhibition of HDAC1) and H3K9 as well as H40K20 methylation (SUV39H1/SUV420H2 inhibition) temporarily influencing the epigenetic pattern. Furthermore, *cgr*-miR-125b-5p could inhibit nuclear translocation of these factors and its complexes as well, since at least MLL1 is associated with WDR5 within cytoplasm prior activation (Wang *et al.* 2012).

Finally, the physiological role of *cgr*KmtTtc36 fusion protein, which is less sensitive on inhibition mediated by inflammatory response, and its role after overexpression of *cgr*Ttc36 need to be elucidated (Figure 5.10). The additional features in *cgr*KmtTtc36 fusion protein are the tetratricopeptide repeat domains providing the binding to heat shock proteins such as *cgr*Hsp70 and other TPR-recognising proteins. Focussing on Hsp70, this protein was found to be variably translocated between nucleus and cytoplasm depending on stress or other signalling (Manzoor *et al.* 2014). Other findings revealed withhold of proteins by Hsp70 within cytoplasm (Gurbuxani *et al.* 2003). Here, overexpression of HSP70 led to reduced apoptosis due to binding of apoptosis-inducing factor (AIF) and predominant maintenance within cytoplasm. In analogy, *cgr*Hsp70 could be localised within nucleus as well as cytoplasm. Thus, *cgr*Hsp70 would be theoretically capable for binding *cgr*KmtTtc36 within nucleus, which for its part would bind H3K4me2/3 residues in transcriptionally active genomic regions. In the following, this binding could recruit histone deacetylases such as *cgr*Hdac1. The human homologue is capable to bind HSP70, which even increase the deacetylase activity (Johnson *et al.* 2002). Furthermore, HDAC1 was reported to interact with the histone demethylase LSD1/KDM1 forming a gene repressor complex (Lee *et al.* 2006; Janzer *et al.* 2012). This interaction would thus lead to H3K4me2/3 demethylation and histone deacetylation, which finally lead to gene repression by additional recruitment of SUV39H1 to HDAC1 and H3K9 methylation after prior LSD1/KDM1 dissociation (Vaute *et al.* 2002; Wang *et al.* 2009a). Therefore, *cgr*KmtTtc36 would putatively act as an anchor for H3K4m2/3 residues in transcriptionally active sites and would facilitate their inactivation (Figure 5.10 A).

In contrast, overexpression of *cgr*Ttc36 would possibly prevent *cgr*Hsp70 translocation within the nucleus and thus initiate Hdac1/Lsd1/Suc39h1 repressor complex binding to *cgr*KmtTtc36 (Figure 5.10 B). Due to the high transcription level of *cgr*Ttc36 (Figure 4.46), it is assumed that *cgr*Hsp70 could be fully maintained within cytoplasm (Chapter 5.3.2.2). The now accessible TPR domains bind other proteins such as p300/*cgr*Kat3b (lysine acetyltransferases 3b, 300 kDa), which was originally documented to bind the TPR motif of TTC5/Strap (Demonacos *et al.* 2001; Lynch *et al.* 2013).

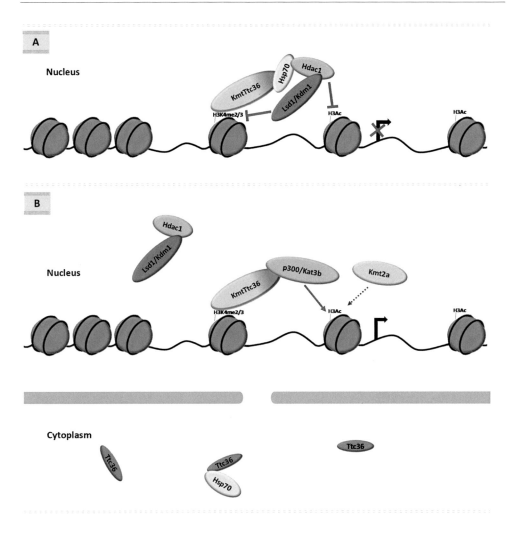

Figure 5.10: Putative mechanism of *cgr*Ttc36 overexpression on *cgr*KmtTtc36-mediated epigenetic regulation. This puristic scheme shows the possible binding partners of *cgr*KmtTtc36 in wild type (A) and *cgr*Ttc36-overexpressing cells (B). In wild type cells, *cgr*KmtTtc36 would putatively act as an anchor recognising H3K4me2/3 and due to Hsp70 binding by its *cgr*Ttc36-TPR domains, *cgr*KmtTtc36 would interact with histone deacetylase Hdac1 as well as histone demethylase Lsd1/Kdm1 leading to gene repression. The overexpression of *cgr*Ttc36 would keep *cgr*Hsp70 within the cytoplasm, whereby *cgr*KmtTtc36 would bind the histone acetylase p300/Kat3 via its *cgr*Ttc36-TPR domains resulting in transcription activation. Noting, this mechanism would be dependent on other factors and tightly regulated towards a histone demethylation/methylation as well as deacetylation/acetylation equilibrium. Other binding partners such as Wdr5 and RbBP5 are not shown due to display clarity.

Thus, the recruitment of p300/*cgr*Kat3b mediates the acetylation of proximal histone residues and further H3K4 methylation by e.g. *cgr*Kmt2a/MLL1 (Bhan *et al.* 2013). Therefore, *cgr*Ttc36 overexpression would result in stabilisation of transcriptionally active genomic regions. All these assumptions were made regarding a present basal expression on *cgr*KmtTtc36, which might be regulated by NF-κB pathway activity that was outlined above. In addition, the putative mechanism would underlay acitivities of other factors and might be tightly regulated towards a histone demethylation/methylation as well as deacetylation/acetylation equilibrium.

Now, the putative DNA-methylation independent inhibitory effects on the *cgr*Ttc36-mediated growth promoting effects first mentioned in chapter 4.3.4.2 are elucidated (Chapter 5.3.2.4).

5.3.2.4 Putative mode of inhibition of *cgr*Ttc36-mediated effects

Next to the competition inhibition by higher affinity or more concentrated Hsp70-binding proteins, other inhibition mechanisms such as structural changes of *cgr*Ttc36 after posttranslational modifications are possible. Epigenetic knockdown of *cgr*Ttc36 mediated by DNA methylation was shown to be not a mode of inhibition in *cgr*Ttc36-overexpressing CHO-K1 cells even though the coding sequence was shown to possess several CpG islands prone for methylation (Figure 4.38). In addition, epigenetic knockdown of *cgr*Ttc36 mediated by histone modifications can be excluded since the mRNA content was maintained after phenotypic alteration (data not shown, chapter 4.3.4.2). Therefore, other inhibitory mechanisms are discussed such as post-translational inhibition of either *cgr*Ttc36 and its associated protein *cgr*Hsp70 or epigenetic effect based on *cgr*KmtTtc36 fusion protein mechanism (Figure 5.10).

Possible inhibitory modifications for *cgr*Ttc36 could be SUMOylation as well as phosphorylation (Chapter 5.3.2.1). At least phosphorylation at certain exposed serine or threonine residues was predicted to change the N-terminus of *cgr*Ttc36 (Figure 5.5) and therefore the binding capacity of the predicted proximate chaperone binding sites (SKALELQ, Table 5.8) by masking or exposure. The likely SUMOylation of the exposed lysine 32 (K32) in *cgr*Ttc36 (Table 5.4) could possess a distinct inhibitory effect of chaperone binding in a similar manner. The reduced *cgr*Hsp70-binding would facilitate *cgr*Hsp70 degradation by *cgr*Chip-mediated ubiquitination (Jiang *et al.* 2001) or would even induce Hsp70-dependent degradation of other proteins such as c-Myc (Paul *et al.* 2012).

Furthermore, *cgr*Ttc36 and *hsa*TTC36 showed motif similarities to lysine methyltransferases modifying Hsp70 and other chaperones as well as tumour suppressors (Wang *et al.* 2011; Jakobsson *et al.* 2013). Additionally, SMYD2 (SET and MYND domain containing 2) was reported to mediate lysine monomethylation of tumour suppressor protein p53 by its tetratricopeptide repeat domain as well as an EDEE sequence (Wang *et al.* 2011). Both elements could be slightly altered found in *hsa*TTC36 and *cgr*Ttc36 (Table 5.10). Interestingly, the crystal structure of SMYD2 shows strong similarities to the predicted *cgr*Ttc36 structure substantiating this hypothesis (Figure 5.3; Jiang *et al.* 2011).

Table 5.10: Putative N-terminal p53 binding motif in *cgr*Ttc36 and *hsa*TTC36 as well as possible inhibition by SUMOylation. The SUMOylation sites are shown in red letters regardless the probability. Putative p53 binding motifs are underlined.

*cgr*Ttc36	*hsa*TTC36
29-EEAKKEDEDGV-39	30-EEAEKEEREEDEVF-43

SUMOylation sites were predicted by GPS SUMO 1.0 (Ren *et al.* 2009) and verified by SUMOplot™.

The EDEE motif in SMYD2 was crucial for binding to p53 and if substitutions by aspartate/glutamate residues would not influence the binding affinity, *cgr*Ttc36 as well as *hsa*TTC36 could harbour this characteristic and mediate lysine methylation in p53 without possessing the relevant enzyme activity but recruiting lysine methyltransferases. Other lysine methyltransferases were even shown to stabilise Hsp70/HSP70 activity (Jakobsson *et al.* 2013), but could not be functionally transferred to *cgr*Ttc36 or *hsa*TTC36. Since the putative p53 binding site in *cgr*Ttc36 (34-EDED-37) was in vicinity to the predicted SUMOylation site (Tables 5.4 and 5.10), it was assumed that a possible p53 destabilisation by *cgr*Ttc36 could be inhibited by SUMOylation, whereat *hsa*TTC36 would not be SUMOylated (Table 5.4). Besides the binding and stabilisation/destabilisation of *cgr*Ttc36 by its TPR domains as well as its predicted SUMO-binding site, this Hsp70-independent mechanism would explain the maintained viability prolongation in thawed 25-CHO-S/2.5C8/*cgr*Ttc36 cell populations compared to the mock transfected control, whilst cell densities were persistently abolished or mock-levelled (data not shown; chapter 4.3.4.2). Nevertheless, the persistent inhibition of the maximal cell density once increased by *cgr*Ttc36 overexpression could not be explained by these rather transient as well as short-termed mechanisms. Thus, because of the reasons above as well as the poor conservation of SUMOylation site in Ttc36/TTC36 evolution and due to the rather transient effect of serine/threonine phosphorylations, the loss in *cgr*Hsp70-binding was suggested to be directly connected with downregulation or inactivation of *cgr*Hsp70. In similarity to SUMOylation, the kinase p38MAPK could putatively induce structural changes in *cgr*Ttc36 by serine 151 phosphorylation (Figure 5.5) in a transient manner. A consistent activation of p38MAPK after DMSO incubation due to freezing-dependent cell banking and subsequent thawing could possibly lead to a lower binding capacity to *cgr*Hsp70. But overall, DMSO was documented to influence recombinant protein expression as well as cell cycle arrest temporarily only (Liu and Chen 2007; Fiore *et al.* 2002), without any significant changes using transiently as well as stably transfected CHO cell lines (Radhakrishnan *et al.* 2008) and putatively by increasing the osmolality that was shown to increase p38MAPK level (Han *et al.* 1994). Therefore, p38MAPK is regarded as a putative transient, post-translational regulator of *cgr*Ttc36.

Cryopreservation of zebrafish (*Danio rerio, dre*) genital ridges was documented to induce epigenetic changes and transcriptional alterations (Riesco *et al.* 2013). Hence, an epigenetic mechanism other than CpG island methylation was suggested, since DNA demethylation by 5-Aza (Chapter 4.3.4.3) did not reveal a phenotypic change in frozen and thawed *cgr*Ttc36-overexpressing 25-CHO-S/2.5C8 cell populations. Therefore, histone acetylation, methylation, demethylation or deacetylation processes are thought to be appropriate to persistently activate or repress *cgr*Ttc36 effect modulators.

The DNA methyltransferase 1o (Dmt1o) was slightly downregulated in vitrified mouse oocytes directly after thawing, whilst expression level steadily recovered over time again (Zhao *et al.* 2013b). In parallel, the Hdac1 histone deacetylase as well as Hat1 histone acetyltransferase transcription levels were not altered by vitrification. In parallel, DNA methyltransferase 1 (dnmt1) expression in the marine flatfish Senegalese sole (Solea senegalensis) increased during low temperature incubation (Campos *et al.* 2013), assuming a species independent effect on DNA methylation. Further, contrary reports showed a significant freezing-dependent downregulation of Hdac1 in mouse oocytes and the resulting embryos after *in vitro* fertilisation (Li *et al.* 2011). A functionally indirect coincidence was reported in vitrified pig oocytes after thawing (Spinaci *et al.* 2012). Here, acetylation of histone 4 (H4) as well as methylated H3K9 content significantly increased two hours after thawing the vitrified pig oocytes, suggesting the delayed but persistent downregulation of Hdac1 resulting both histone acetylation and methylation (Figure 5.10). Simultaneously, the histone acetyltransferase p300/KAT3B was shown to be increased in H1299 cells after lowering of the temperature from 37 °C to 32 °C, whilst parallel upregulation of MDM2 as well as downregulation of p53 (Kawai 2001). Downregulation of p53 by MDM2 by shifting to low temperature was observed in the BALB/3T3 mouse cell line, suggesting a similar mechanism (Sakurai *et al.* 2005). Complementary, the freezing agent dimethylsulfoxide (DMSO) was reported to activate the cytomegalovirus (CMV) promoter (P_{CMV}) in CHO-K1 cells by increasing the level of acetylated histone H3 and activation of p38 mitogen-activated protein kinase (p38MAPK) as well as nuclear factor κB (NF-κB) signalling pathways (Radhakrishnan *et al.* 2008). A 6-fold activation of NFκB signalling pathways was observed in human umbilical vein endothelial cells (HUVEC) by mild hypothermic conditions as well (Roberts *et al.* 2002), whilst the NF-κB pathway protein p65/RelA was shown to bind the histone acetylase p300/KAT3B and inhibit the p53 transactivation in HEK293 as well as U2OS cells (Webster and Perkins 1999). In similar, the inhibition of histone deacetylases (HDACs) through sodium butyrate (NaBu) application resulted in increased H3K4 trimethylation possibly mediated by Kmt2a in mice (Gupta *et al.* 2010).

Finally, HSP70 as well as HSP90 were downregulated during cryopreservation properties (high DMSO concentration and low temperature) in HeLa cervix adenocarcinoma cell line, whereat thawing lead to a successive increase in HSP70/90 expression (Wang *et al.* 2005). The same result of transient increase of HSP70 expression (4-fold, eight hours after culture temperature retrieval) was achieved using a different experimental design (acclimation at 25°C, warming to 37 °C) and another human cell line (WI26) (Neutelings *et al.* 2013). Moreover, the upregulation of Hsp70.1 negatively regulated the NF-κB pathway, resulting a reset of DMSO/hypothermic-driven signalling cascade and restorage of the initial state (Kaneko and Kibayashi 2012).

Taken these findings together, the conditions during cryopreservation (-1°C/minute, 10 % DMSO; chapter 3.4) might result in epigenetic changes due to upregulation or stimulation of H3K4 methylases, histone acetyltransferases as well as downregulation of repressive epigenetic modulators such as histone deacetylases. Suggesting an identical mechanism in *cgr*Ttc36-overexpressing as well as control CHO-S cell lines and regarding the main differences in *cgr*KmtTtc36-directed mechanism (Figure 5.10), the ability of the putative *cgr*KmtTtc36-p300 complex in *cgr*Ttc36-overexpressing cells would lead to an even increased enhancer/promoter activation during cryopreservation exemplarily due to subsequent histone deacetylase downregulation. In contrast, the extent of persistent epigenetic enhancer/promoter activation would be lower in wild type cells due to the initially repressive state of *cgr*KmtTtc36-mediated complexes. Nevertheless, downregulation of *cgr*KmtTtc36 by NF-κB signalling would putatively proceed earlier than downregulation of *cgr*Kmt2a due to prior NF-κB-mediated miR-125b-5p maturation, possibly leading to increased *de novo* H3K4 methylation as well as histone acetylation. This sustained activation mechanism would be available in wild type cells but at lower extent. Therefore, this fundamental mechanistic distinction would putatively conduct various inhibitory gene activations in *cgr*Ttc36-overexpressing cells resulting in an abolishment in *cgr*Ttc36-mediated action.

A putative mechanism could also be the sustainably higher activity of RelA hetero-/homodimers or alternative splicing modulators after cryopreservation, since the CV146-derived IntronA-*cgr*Ttc36-IRES$_{FMDV}$-HygroR mRNA possibly possesses two alternative splice sites either within the *cgr*Ttc36 coding sequence or the internal ribosomal entry site (IRES$_{FMDV}$) (Figure 5.11). Moreover, a putative NFκB1/RelA binding site was predicted in proximity of the IntronA acceptor splice site, presumably leading to an inhibition of splicing at this acceptor site after NF-κB pathway activation through direct mRNA-binding (Khan *et al.* 2012).

Interestingly, binding sites of documented modulators of alternative splicing were predicted on the IntronA-*cgr*Ttc36-IRES$_{FMDV}$-Hygro^R mRNA (Figure 5.11): the RNA binding motif proteins 3 (RBM3/Rbm3) and CIRBP/Cirbp (cold inducible RNA binding protein) as well as the small nucleolar C/D box RNA Snord115/MBII-52. RBM3/Rbm3, which expression was reported to mediate and increase alternative splicing in CD44 cell-surface glycoprotein through its RNA recognition motif (RRM), could act as a possible alternative splice modulator (Zeng *et al.* 2013). In companionship with the RRM-containing CIRBP/Cirbp, RBM3 is expressed during incubation at low temperatures (Shin *et al.* 2011; Kaneko and Kibayashi 2012), whereat the expression level of RBM3 was reported to be maintained longer than CIRBP after warming to 37 °C (Neutelings *et al.* 2013). CIRBP on the other hand was shown to increase the RelA (p65) as well as NFκB1 (p50) protein levels of the NF-κB pathway (Brochu *et al.* 2013), which could additionally induce alternative splicing. At least in mice, Cirbp as well as Rbm3 are capable to induce alternative polyadenylation upon cold shock, suggesting subsequent mRNA stabilisation (Liu *et al.* 2013b). In case of the IntronA-*cgr*Ttc36-IRES$_{FMDV}$-Hygro^R mRNA, alternative polyadenylation would putatively occur after the second AAUAAA within the SV40 polyA signal (Table 3.20).

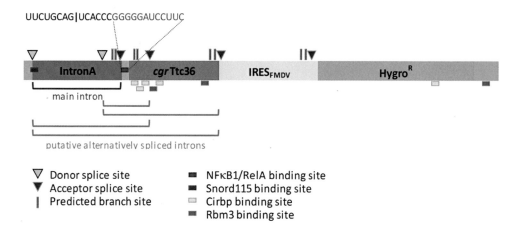

UUCUGCAG|UCACCCGGGGGAUCCUUC

Figure 5.11: Schematic overview of possible splice sites in CV146-derived mRNA including putative NFκB1/RelA, Snord115, Cirbp and Rbm3 binding sites. The NFκB1/RelA (p50/p65) binding sequences (either GGGGGAUCCUU or GGGGAUCUUC) were predicted using recent consensus sequence optimisation (Wong *et al.* 2011) and are located in vicinity of the main IntronA splice site. The putative Snord115 binding site overlaps with IntronA donor splice site. The vast majority of the predicted Cirbp ([U/G]G[G/A]AG[A/C][A/U]G) as well as Rbm3 binding sites (G[C/G][U/A]G[G/C][A/U]G) (Liu *et al.* 2013b) are located within *cgr*Ttc36 coding sequence. Alternative splice sites are situated within the coding sequence of *cgr*Ttc36 or within the IRES$_{FMDV}$ (internal ribosomal entry site derived from foot-and-mouth disease virus) resulting *cgr*Ttc36 non-sense splicing.

Downregulation of RBM3 was shown to deregulate the maturation and expression of various pre-miRNA species (Pilotte *et al.* 2011). Therefore, it was suggested that RBM3 would influence splicing of *cgr*Ttc36 expression cassette in frozen/thawed 25-CHO-S/2.5C8/*cgr*Ttc36 directly or mediated by miRNA-modulation (Pilotte *et al.* 2011; Dresios *et al.* 2005). Some miRNA species were shown to have the capability to mediate alternative splicing mechanisms (Wu *et al.* 2013), but no miRNA could be proven to be involved in alternative splicing of IntronA-*cgr*Ttc36-IRES$_{FMDV}$-Hygro^R mRNA. Nevertheless, a C/D box small nucleolar RNA (snoRNA) Snord115/MBII-52 binding site was predicted at the IntronA donor splice site (Figures 5.11 and 5.12). This snoRNA was shown to mediate alternative splicing on various targets (Kishore *et al.* 2010) and downregulation led to massive alterations in splice product content (Doe *et al.* 2009). The alignment of *mmu*Snord115 (gi|14277022|gb|AF357427.1) with CHO-K1 unplaced genomic scaffold scaffold2613 (gi|344162245|gb|JH002820.1|) revealed the homologues *C. griseus* Snord115 sequence (*cgr*Snord115) (Figure 5.12).

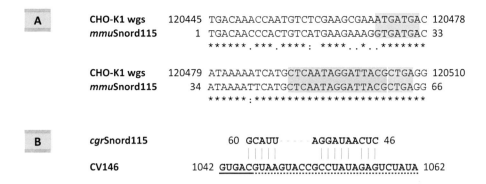

Figure 5.12: Genomic sequence alignment of *cgr*Snord115 (A) and its putative binding site in IntronA donor splice site (B). The *C. griseus* sequence for Snord115 (*cgr*Snord115) was revealed by alignment of *mmu*Snord115 (gi|14277022|gb|AF357427.1) with CHO-K1 unplaced genomic scaffold scaffold2613, whole genome shotgun sequence (wgs) (gi|344162245|gb|JH002820.1|) (A). The C'-box ([A/G]TGATGA) as well as D-box of both Snord115 species are shaded in darker red, whereat putative antisense box (CTCAATAGGATTACG) is shaded in lighter red (regarding Kishore *et al.* 2010). The putative *cgr*Snord115 binding site in CV146-IntronA overlaps with donor splice site (exon/intron) (B).

The antisense box within *cgr*Snord115 was found to be highly homologous and revealed high coverage against IntronA donor splice site. Therefore, it is assumed that the transcription of *cgr*Snord115 would lead to an alternative splice pattern in CV146-containing cells. Due to this observation, it is further hypothesised that the transcriptional activating complex (*cgr*KmtTtc36/p300; Figure 5.10) would induce a persistent activation of e.g. *cgr*Cirbp, *cgr*Rbm3, *cgr*Snord115 and/or the NF-κB pathway proteins after hypothermic/DMSO treatment leading to a subsequent nonsense splicing of the IntronA-*cgr*Ttc36-IRES$_{FMDV}$-Hygro^R mRNA. The latter mechanism could not be distinctively observed with present qRT-PCR primer pair (Table 3.8, Ttc36_fw/rv) because of the absence of intronic sequence overlap. Hence, the fade of *cgr*Ttc36, *cgr*KmtTtc36 and their modulators needs to be monitored for further studies based on mRNA as well as protein level elucidating the role of *cgr*Ttc36 in its epigenetic regulation and the role of phenotypic change after hypothermic as well as DMSO treatment. Hereinafter, possible optimisations could be performed primarily regarding the sequence (reduced CpG-content, point mutations avoiding e.g. SUMOylation, abolishment of alternative splice or effector binding sites and replacement of IRES$_{FMDV}$ sequence). The hypothesises above would give a first view on possible actions and inhibitory pathways leading to the observed abolishment of *cgr*Ttc36-mediated effects. In following, the TPR-containing protein *cgr*C10orf93-like/*cgr*Ttc40-like is discussed regarding the possible structure and action (Chapter 5.3.3). In addition the overexpression of *cgr*C10orf93-like/*cgr*Ttc40-like showed similar aggregation inhibition such in *cgr*Ttc36-overexpressing cell lines suggesting a common mechanism (Figures 4.25 and 4.41).

5.3.3 The putative localisation as well as function of *cgr*C10orf93-like/*cgr*Ttc40-like

The tetratricopeptide repeat (TPR)-containing protein *cgr*C10orf93-like/*cgr*Ttc40-like is functionally unknown due to absence of reported binding partners or comparable studies with homologues protein species. Therefore, further discussions on the effect of overexpression in various CHO-K1 derived cell lines tend to be limited (Chapter 4.3.2). Nevertheless, some functional motifs and regulatory elements should subsequently be revealed to estimate the putative cause of *cgr*C10orf93-like/*cgr*Ttc40-like-driven growth inhibition as well as increase in specific productivity. For better understanding, the putative structure, motifs, binding as well as modification sites were predicted (Figure 5.13).

The cellular localisation of *cgr*C10orf93-like/*cgr*Ttc40-like was predicted to be in cytoplasm predominantly, whereat two weak nuclear localisation sequences (NLS) and two export signals (NES) were predicted (Figure 5.13; Table 5.11). Further, the localisation prediction by NucPred (Brameier *et al.* 2007) revealed a score of 0.85 that represent a predominant cytosolic localisation. Therefore, *cgr*C10orf93-like/*cgr*Ttc40-like is a predictively exclusive cytosolic protein with various predicted binding domains.

Next, a sole TPR domain was located between residues 468 – 501, whereat a similar secondary structure such observed in *cgr*Ttc36 was predicted by NetSurfP (Petersen *et al.* 2009) as well as NetTurnP (Petersen *et al.* 2010) (Figure 5.2). In comparison with *cgr*Ttc36, the putative chaperone binding site within the single TPR domain of the 126 kDa protein *cgr*C10orf93-like/*cgr*Ttc40-like possesses high homology to other TPR-containing species and an enhanced positive current due to an additional arginine (Table 5.12). The latter would possibly increase the binding to the lipophilic as well as negatively charged C-terminal (M/I)EEVD and PTIEEVD peptides of heat shock proteins as well as chaperones such as Hsp70/HSP70 (Matsumura *et al.* 2013). The second chaperone binding site predicted in *cgr*Chip, *cgr*TPRp and *cgr*Ttc36 could not be retrieved in *cgr*C10orf93-like/*cgr*Ttc40-like also because of the presence of a single TPR domain only. Furthermore, the *cgr*C10orf93-like/*cgr*Ttc40-like TPR domain shows significant coverage with a predicted nebulin domain (position 496 – 512) (Figure 5.13) suggesting a competitive interaction mechanism resting upon the overlapping sequence including the putative chaperone binding site. The appearance of this putative nebulin domain might enable *cgr*C10orf93-like/*cgr*Ttc40-like for its binding to filamentous actin (Wang *et al.* 1996; Moncman and Wang 1995) further assuming the involvement in focal adhesion regulation (Panaviene and Moncman 2007). Due to the nebulin/TPR domain overlap, the expression of heat shock proteins could lead to a dissociation from filamentous actin and disruption or even activation of its adhesion-dependent action by competition (Deng *et al.* 2008; Bliss *et al.* 2013). The tetratricopeptide repeat domain 9A (TTC9A) was shown to be involved in actin filament regulation by its TPR domains and overexpression leads to an increased cell mobility by stabilisation of tropomyosin (Cao *et al.* 2008; Cao *et al.* 2006). This illustrates the potential of tetratricopeptide repeat-containg proteins in focal adhesion regulation even in inducible focal adhesion rearrangements (Shrestha *et al.* 2012).

Figure 5.13: The predicted secondary structure of *cgr*C10orf93-like/*cgr*Ttc40-like. Following features were predicted by NetSurfP (Petersen *et al.* 2009) as well as NetTurnP (Petersen *et al.* 2010) and highlighted as following: α-helices (red bars), β-turns (faint red bars), β-strands (orange bars) as well as coil structures (grey bars). Bars were manually generated using the relative probability values (> 0.500) of raw data output. The highlighted motifs (TPR: tetratricopeptide repeat; Nebulin: nebulin domain; P-loop: ATP/GTP-binding site motif A; FATC: FRAP, ATM, TRRAP C-terminal domain) were predicted by PROSITE as well as Pfam search (Sigrist *et al.* 2009; Pagni 2001). The other motifs (NLS: nuclear localisation sequence; NES: nuclear export signal; SUMO: SUMO-binding site) are displayed using predictions in tables 5.11 and 5.13.

Table 5.11: Putative as well as predicted motifs and modification sites in *cgr*C10orf93-like/*cgr*Ttc40-like.

Motive	Sequence	Position	Score
Nuclear export sequences (NES)[†]	LLFELGRLSL	291-300	0.65
	LECELEALRL	332-341	0.79
Nuclear localisation sequences (NLS)[‡]	INSLKAKLDKNDLPEDIDQILMNAFKHLSH	246-276	5.3
	PVNRKKAKGS	1090-1099	5.5

[†] Nuclear export sequences were predicted using NESsential (Fu *et al.* 2011) and verified with reported consensus sequences (La Cour *et al.* 2004) (Scores >0.60 only). The scores represent the probabilities indicating the likelihood of a functional NES or the occurrence of a NES (In the range from 0 to 1).

[‡] Nuclear localisation sequence prediction by cNLS Mapper (Kosugi *et al.* 2009). Scores represent the localisation probability in cytoplasm or nucleus, whereat scores of 3 – 6 indicate localisation to both the nucleus and the cytoplasm.

147

Furthermore, a FATC (FRAP, ATM, TRRAP C-terminal) domain (position 699-752) was predicted, which putatively allows binding of *cgr*C10orf93-like/*cgr*Ttc40-like to plasma membrane (Sommer *et al.* 2013). This observation substantiates the attendance of *cgr*C10orf93-like/*cgr*Ttc40-like in regions with proximal filamentous actin as well as lipid membranes such as focal adhesions (Ciobanasu *et al.* 2013)). Interestingly, the appearance of a predictive ATP/GTP-binding site motif A (P-loop) at position 576-583 (AKVSTGKT; consensus: [G/A]X$_4$GK[T/S]) reveals the possibility to bind ATP or GTP and the possible usage for e.g. chaperone mechanisms analogous to *cgr*Hsp70, which was found to possess a P-loop for crucially mediating its chaperone function (Lewis and Pelham 1985). Further mechanism could be the negative regulation by GTPase activating proteins leading to an inactivation of *cgr*C10orf93-like/*cgr*Ttc40-like (Nobes and Hall 1995).

Table 5.12: Chaperone binding site alignment of *cgr*C10orf93-like/*cgr*Ttc40-like with *cgr*Ttc36, *cgr*TPRp (both experimental sequences) and *cgr*Chip (XP_003501570.1).

*cgr*Ttc40-like	492-KKAMRLD-498	850-KR-NVRPNLVHIPME-863
*cgr*Chip	113-RRALELD-119	64-RR-QLLP----IPRE-73
*cgr*TPRp	5-KKxFxIx-11	96-RKxDLxx---xxPxx-107
*cgr*Ttc36	48-SKALELQ-54	155-RR-QLVL---LNPY-164
	:.: :	:: :: *

Another function could be the binding and signal transduction modulation of SUMOylated proteins, since two SUMO-binding sites were predicted (Table 5.13). Nevertheless, overexpression of *cgr*C10orf93-like/*cgr*Ttc40-like rather revealed growth inhibition as well as early growth arrest but led to a slightly increased specific productivity (Figures 4.23 and 4.24).

Table 5.13: Putative as well as predicted motifs and modification sites in *cgr*C10orf93-like/*cgr*Ttc40-like.

Motive	Sequence	Position	Score
SUMO-binding	IPSLSQIITVLNQTEEDKE	160 – 164	64.86
	RLGDPTAIHVLCTTQWNTC	379 – 383	64.34
	LMDDNQLKEEKKHSA	231	27.89
SUMOylation	HQRFPSVKEEKMLLL	285	43.78
	GRLSLILKNESMASD	305	25.52

SUMOylation and SUMO binding sites were predicted by GPS SUMO 1.0 (Ren *et al.* 2009). The type-I consensus sequence (Ψ-K-X-E) is underlined. Thresholds were set to 19.71 (SUMOylation) and 59.29 (SUMO-binding). The consensus as well as modified residues are shown in red letters.

Therefore, the overexpression of *cgr*C10orf93-like/*cgr*Ttc40-like (63-fold, data not shown) could inhibit viable cell mechanisms. Due to the adherence independence of the suspension CHO-K1 cell lines used, focal adhesions are not likely to occur. Therefore, this mechanism is suggested to be negotiable. On the other hand, *cgr*C10orf93-like/*cgr*Ttc40-like could positively affect adherent cell lines due to the presence of cell-matrix interactions such as focal adhesions. Generally, it is hypothesised that the overexpression of *cgr*C10orf93-like/*cgr*Ttc40-like would inhibit cell-cell aggregation by impairing the adherence mechanisms by either active inhibition or concentration dependent sponge-like quenching of aggregation stimuli. This mechanism should be further elucidated in chapter 5.3.4.

The native physiologic function of *cgr*C10orf93-like/*cgr*Ttc40-like could be regulated by three possible SUMOylation sites, which could influence the cellular localisation, due to their proximity to the two nuclear export signals as well as the bipartite nuclear localisation sequence (Figure 5.13). Moreover, *cgr*C10orf93-like/*cgr*Ttc40-like could be regulated by phosphorylation as well as by O-linked glycosylation (O-GlcNAc) (Table 5.14). For the latter modification, four O-GlcNAcylation sites (T80, S448, S864 and T997; scores > 0.50; exposed residues underlined) were found using YinOYang 1.2 prediction tool (Gupta and Brunak 2002). Here, no predicted motif or phosphorylation site was predicted to be targeted. The protein kinase C (PKC) as well as the cAMP-dependent protein kinase (PKA) were highly predicted for mediating the phosphorylation of buried and exposed serine/threonine residues of *cgr*C10orf93-like/*cgr*Ttc40-like. Applying a threshold >0.75 on exposed residues only revealed the sites in table 5.14.

Table 5.14: Predicted phosphorylation sites in *cgr*C10orf93-like/*cgr*Ttc40-like located in regulatory features or probably performed by known kinases.

Exposed serine/threonine residues[†]	Predicted kinase[‡]	Target motif
248	**PKC**	**NLS**
283	PKC	-
343	**PKA**	**(NES)**
437	PKC	-
579	-	**P-loop**
702	-	**FATC**
796	PKC	-

† Exposed serine/threonine residues were predicted by NetSurfP (Petersen *et al.* 2009) and phosphorylation sites by NetPhos 2.0 (Blom *et al.* 1999) (Scores > 0.75).
‡ Predicted by NetPhosK 1.0 (Blom *et al.* 2004) (Scores > 0.75). PKA: Protein kinase A, cAMP-dependent protein kinase; PKC: Protein kinase C.

Here, the bipartite NLS, the P-loop (ATP/GTP-binding site motif A) and the FATC (FRAP, ATM, TRRAP C-terminal) domain were predicted to possess the ability to be phosphorylated directly, whereat the second NES could be influenced by the proximal phosphorylation site. Due to possible structural changes analogous to *cgr*Ttc36 (Figure 5.5), other motifs and binding domains could be influenced by phosphorylation. Hence, this highly regulated multi-functional protein needs further investigations, gaining knowledge for subsequent sequence optimisation or leading to alternative ways for its use in biopharmaceutical protein production. The present wildtype *cgr*C10orf93-like/*cgr*Ttc40-like is not capable for increasing the cell specific productivity. Nevertheless, applying fragmental *cgr*C10orf93-like/*cgr*Ttc40-like polypeptides to CHO-K1 derived cell lines would eventually lead to improved producer cell lines. Especially the beneficial inhibition of aggregation could be of interest. Therefore and due to findings in *cgr*Ttc36-overexpressed cell lines, the mechanism behind this inhibition in cell aggregation is further elucidated (Chapter 5.3.4).

5.3.4 The inhibition of cell aggregation in *cgr*Ttc36 and *cgr*C10orf93-like/*cgr*Ttc40-like overexpressing cell lines

Both, *cgr*Ttc36 as well as *cgr*C10orf93-like/*cgr*Ttc40-like were shown to decrease the grade of aggregation in K20-3-derived cell lines in a similar manner (Figures 4.25 and 4.41). Further, clonal variations in K20-3/*cgr*Ttc36 revealed a concentration independent mechanism suggesting a strong inhibition of cell-cell adhesion participating proteins (Figure 4.45). In contrast to the common anti-aggregation effect as well as containment of tetratricopeptide repeat (TPR) domains, both proteins showed no further structural or phenotype-modulating similarities. The 126 kDa protein *cgr*C10orf93-like/*cgr*Ttc40-like was shown to possess various regulatory domains such as an ATP/GTP-binding P-loop as well as nebulin domains (Chapter 5.3.3). On the other hand, the much smaller 20.4 kDa protein *cgr*Ttc36 was predicted to possess three tightly alternating TPR domains with no additional regulatory domains incorporated within the adherens junction regulation pathways (Chapter 5.3.2).

Common motifs in *cgr*Ttc36 as well as *cgr*C10orf93-like/*cgr*Ttc40-like are the tetratricopeptide repeat (TPR) motifs, whereat *cgr*Ttc36 was found to possess three homologous motifs and *cgr*C10orf93-like/*cgr*Ttc40-like simply one TPR motif (position 468 – 501) instead (Table 5.15). Therefore, efficient chaperone binding by *cgr*C10orf93-like/*cgr*Ttc40-like could not be performed, since three serial tetratricopeptide repeat motifs are crucial for ligand binding (Petters *et al.* 2013). Possibly due to the long history of CHO-K1 in *in vitro* culture, the first TPR motif (TPRα) in *cgr*Chip was evolutionary abolished and showed poor homology to *hsa*CHIP (NP_005852.2) as well as *mmu*Chip (NP_062693.1), which TPR region possesses of the minimal three TPR motifs (Ballinger *et al.* 1999). Therefore, *cgr*Chip would not be capable to bind *cgr*Hsp70, which could be adapted to the mono-TPR containing protein *cgr*C10orf93-like/*cgr*Ttc40-like. The human homologue TTC40 (NP_001186978.2) was predicted to possess two spatially distinct TPR motifs (positions 426 – 459 as well as 1111 – 1144) (predicted by PROSITE; Sigrist *et al.* 2009). In further studies, the binding capability of these CHO-K1 derived proteins is worth for elucidation. In contrast, the TPR region (TPRα-TPRβ-TPRγ) of *cgr*Ttc36 revealed a strong relationship to homologues TPR regions observed in e.g. *cgr*Chip (Carboxyl-terminus of Hsp70/Hsc70 interacting protein; XP_003501570.1; position 89 – 156; TPRβ-TPRγ)) as well as *cgr*Stip1 (stress-induced-phosphoprotein 1; NP_001233607.1; position 4 – 105; TPRα-TPRβ-TPRγ; Chen 1998) (alignment not shown). Especially, the TPRβ motif of these species showed the highest homology implicating, a strong structural as well as functional conservation of this flanked TPR motif (Table 5.16). These observations are of increased interest, since heat shock proteins (Hsp) were shown to directly influence appearance and strength of adherens junctions as well as focal adhesions (Guo 2011; Chatterjee *et al.* 2008; Mao *et al.* 2003).

Table 5.15: Motif alignment of *cgr*C10orf93-like/*cgr*Ttc40-like (*cgr*Ttc40l) and *cgr*Ttc36 tetratricopeptide repeat (TPR) domains. Homologous as well as similar residues are shaded in red and grey, respectively.

	TPR domain	Alignment
*cgr*Ttc40l	-	468 CQVHMEMACIEQDEDRLEPAIEHLKKAMRLDSQG 501
*cgr*Ttc36	α	48 SKALELQGVRAAEAGDLHTALEKFGQAICLLPER 81
	β	82 ASAYNNRAQARRLQGDVAGALEDLERAVTLSGGQ 115
	γ	120 RQSFVQRGLLARLQGRDDDARRDFEQAARLGSPF 153

Table 5.16: Tetratricopeptide repeat motif alignment of *cgr*Ttc36 (XM_003510512.1), *cgr*C10orf93-like/*cgr*Ttc40-like (*cgr*Ttc40l; XP_003510630.1), *cgr*Stip1 (NP_001233607.1) and *cgr*Chip (XP_003501570.1). Homologous as well as functionally similar residues are shaded in red and grey, respectively.

	TPR domain	Alignment
*cgr*Ttc36	TPRβ	82 ASAYNNRAQARRLQGDVAGALEDLERAVTLSGGQ 115
*cgr*Ttc40l	TPR	468 CQVHMEMACIEQDEDRLEPAIEHLKKAMRLDSQG 501
*cgr*Chip	TPRβ	89 AVYYTNRALCYLKMQQPEQALADCRRALELDGQS 122
*cgr*Stip1	TPR1β	38 HVLYSNRSAAYAKKGDYQKAYEDGCKTVDLKPDW 71

At least human TTC36 was proven to interact with HSP70 (Liu *et al.* 2008) and additionally HSP90/Hsp90 as well due to C-terminal sequence similarities. This might occur at lower extent because of observed higher dissociation constants for Sti1/Hop and Hsp90 interaction as well as the higher similarity of *cgr*Ttc36 TPR region to inhibitory TPR region in Sti1/Hop (TPR1, partly shown in table 5.16) (Scheufler *et al.* 2000). Furthermore, it was previously suggested that *cgr*Ttc36 would compete with *cgr*Chip leading to stabilisation as well as increase in *cgr*Hsp70 concentration (Chapter 5.3.2.2). The latter suggestion is valid if *cgr*Chip is capable in binding of *cgr*Hsp70, even the minimal TPR region consisting of three serial TPR motifs could not be predicted. Overexpression of *cgr*Hsp70 in CHO-K1 cells revealed a similar phenotype as observed in 25-CHO-S/2.5C8/*cgr*Ttc36 cell line (Figure 4.37), whereat the cell aggregation was not documented in this study suggesting a respectively unaltered phenotype (Lee *et al.* 2009). In addition, Hsp70 or Hsp90 overexpression was reported to result in alternative secretion via endosomal/lysosomal transport thereof (Luo *et al.* 2013b; Evdokimovskaya *et al.* 2010; Mambula *et al.* 2007). By the assumed *cgr*Ttc36-mediated *cgr*Hsp70 stabilisation, the latter could be enriched and subsequently secreted to mediate proliferative mechanisms and reduction of E-cadherin level (Juhasz *et al.* 2013; Li *et al.* 2013a), which importantly mediates adherens junctions and cell-cell adhesion therefore (Canel *et al.* 2013). Further, the possible secretion of *cgr*Hsp70 could accompany with the co-secretion of *cgr*Ttc36, which was observed for HSP90α and associated AKR1B10 (Aldo-Keto Reductase 1B10) (Luo *et al.* 2013b). In this connection, intracellular as well as extracellular Hsp90/90α was reported to crucially mediate cell invasion in several cell types and lines (Sidera *et al.* 2008; Taiyab and Rao 2011). Coherently, the inhibition of intracellular HSP90/Hsp90 ATPase activity or by interception of extracellular HSP90 was shown to massively impair cell-cell adhesion mechanisms and led to an increased expression of HSP70 (Taiyab and Rao 2011; Stellas *et al.* 2010). Hence, *cgr*Ttc36 could interfere the lysosomal alternative secretion pathway by withholding and masking *cgr*Hsp90 within cytoplasm or by inhibition of its extracellular binding ability. Therefore, the excess of *cgr*Ttc36 could lead to an abolishment of the binding of *cgr*Hsp90 by various C-terminal binding partners such as *cgr*Stip1 (Hop/HOP) and mitochondrial transmembrane transport proteins (e.g. import receptor Tom70) (Scheufler *et al.* 2000; Young *et al.* 2003). In addition, *cgr*Ttc36 could facilitate degradation of focal adhesion kinase (Fak/FAK) by masking the C-terminus of Hsp72 (EEVD) (Mao *et al.* 2003), which possesses high similarities to *cgr*Hsp70. This interaction would lead to a reduction of integrin-mediated signal transduction and cell-matrix adhesion (Mao 2004). Moreover, at least the inhibition of Hsp70/CHIP-mediated ubiquitination of Hsp70-binding proteins could lead to massively altered proteome (Stankiewicz *et al.* 2010). Therefore, the overexpression of *cgr*Ttc36 and the resulting deduction of *cgr*Hsp70 and *cgr*Hsp90 from a pool of direct as well as proximal interactors would cause cell-cell attachment independent phenotypes. Nevertheless, the inhibition of cell aggregation by simple overexpression or upregulation of Hsp70/HSP70 could not be found in several studies concerning suspension cell lines (Komarova *et al.* 2004; Khoei *et al.* 2004). Moreover, a previous study concerning CHO cells and changes in adherence behaviour upon heat-shock response after short-termed incubation at 45°C (10 min) revealed either a slight transient detachment of adherent CHO cells or a reduced attachment of prior suspended cells (Cress *et al.* 1990). Subsequently, the heat-treated CHO cells quickly recovered towards complete adherence. The appearance of Hsp70 could not be detected even other studies revealed a very fast expression of Hsp70 in CHO cells upon heat stress under given properties (Cates *et al.* 2011). On the other hand, Hsp70 expression was shown to last up to 48 hours after various heat shock modes applied on CHO cells, i.e. expression would last longer than the time for proper adhesion (Fornace *et al.* 1989). On the contrary, *cgr*Hsp70-*cgr*Ttc36 interaction would lead to a stabilisation of E-cadherin co-receptor c-Met by *cgr*Chip-dissociation (if active in CHO-K1) leading to adherens junction stabilisation (Jang *et al.* 2011). Hence, *cgr*Hsp70-*cgr*Ttc36 interaction and the hypothesised resulting *cgr*Hsp70 excess could lead to inhibition of cell aggregation, but putative other mechanisms could facilitate and support this mechanism. Therefore, other mechanisms need to be mandatory for the observed phenotype(s), especially regarding the clonal differences compared to the K20-3/*cgr*Ttc36 pool cell line (Figures 4.41 and 4.45).

Next, the putative mode of action of *cgr*C10orf93-like/*cgr*Ttc40-like should be examined. Here, the protein is widely unknown in its function but was predicted to possess several motifs such as single TPR, nebulin and FATC motifs with rather low probability as well as a more likely P-loop ATP/GTP binding motif (Figure 5.13). As discussed above, the TPR motif would not be sufficient for binding of e.g. *cgr*Hsp70 or *cgr*Hsp90. Nebulin repeats were observed in actin modulating proteins mediating focal adhesions (Pappas *et al.* 2011). Here, the scaffold protein nebulin predominantly interacts with filamentous actin associated muscle stress fibres and possesses a vast number of nebulin repeats, whereat the smaller nebulin repeat-containing proteins Lasp-1 and Lasp-2 were shown to be associated to focal adhesions as well. Lasp-2 interacts with the focal adhesion scaffold proteins vinculin as well as paxillin, whilst this association is predominantely mediated by its SH3 (Src homology domain 3) domain (Bliss *et al.* 2013). Other publications rather lead to the suggestion that the three nebulin motifs in Lasp-2 are mandatory for focal adhesion translocation as well as actin binding, but possibly could stabilise the binding (Nakagawa *et al.* 2009). The sole nebulin repeat of *cgr*C10orf93-like/*cgr*Ttc40-like would therefore possess no direct function in focal adhesion regulation.

In addition, FATC (FRAP, ATM, TRRAP C-terminal) domains were shown to be associated with membrane-binding (Dames 2010) as well as protein-interactions such with the histone acetyltransferase TRAP60 (Jiang *et al.* 2006). Since *cgr*C10orf93-like/*cgr*Ttc40-like possesses two low-quality nuclear localisation sequences (NLS) and one predicted FATC domain, which could theoretically be regulated by phosphorylation (Table 5.14) the putative nuclear translocation could be mediated upon certain stimuli (e.g. protein kinase C activation or vice versa) and alterations in transcription by binding of e.g. *cgr*Trap60. The third and the most likely motif predicted represents a ATP/GTP-binding P-loop (consensus: [G/A]XXXXGK[T/S]; AKVSTGKT), which was observed in several kinases, G-proteins, heat-shock proteins as well as elongation factors (Saraste *et al.* 1990). Due to the absence of catalytic sequences, *cgr*C10orf93-like/*cgr*Ttc40-like might act as a subunit of a regulatory protein complex either localised in cytosol, attached to the plasma membrane or within the nucleus providing the ATP or GTP binding and a scaffold function. On the other hand, former catalytic functions could be distinguished during protein evolution and putatively led to the observed inhibitory effects upon overexpression. Underlining this suggestion, *cgr*C10orf93-like/*cgr*Ttc40-like possesses regions with several positively charged residues assuming a special role in protein-protein interaction. Furthermore, Ezrin/Radixin/Moesin (ERM) proteins were shown to bind cytosolic K/R triplets of cell adhesion molecules such as CD44 or ICAM-2 (intercellular adhesion molecule 2) (Yonemura *et al.* 1998). The signal transduction of hamster homologues *cgr*CD44 (RRRCGQKKK; XP_003497494.1 – XP_003497501.1) or *cgr*Icam-2 (HWHRRRTGT; XP_003501928.1) leading to cell-cell adhesion could therefore be inhibited by overexpression of *cgr*C10orf93-like/*cgr*Ttc40-like (e.g. LKRVKKKKG or KDSARKKRA; XP_003510630.1) and the masking of binding proteins such as ERM. This effect was reported for transmembrane Neutral endopeptidase 24.11 (Neprilysin), which possesses a lysine-triplet at its N-terminal cytoplasmic residue (*cgr*Neprilysin-like: PKPKKKQRW; ERE89934.1) proven for binding ERM proteins (Iwase *et al.* 2004). Therefore, this feature of *cgr*C10orf93-like/*cgr*Ttc40-like in combination with its membrane-binding FATC domain could influence adhesion proteins directly. Using the InterPro protein sequence analysis and classification tool (Hunter *et al.* 2011), a likely association to the alpha-2 macroglobulin receptor (low-density lipoprotein receptor-related protein 1, Lrp1/LRP1) (1st: GO:0005515, 2nd: GO:0045308) was predicted for *cgr*C10orf93-like/*cgr*Ttc40-like. Coherently, LPR1 was found to interact e.g. with the adhesion-associated transmembrane protein CD44, which could putatively interact with *cgr*C10orf93-like/*cgr*Ttc40-like as well (Perrot *et al.* 2012). Here, downregulation of LPR1 or treatment with its binding partner LRPAP1 (LRP-associated protein 1; also: RAP, receptor-associated protein) led to a stabilisation as well as accumulation of CD44 and therefore to an increased adhesion. On the other hand, LPR1 mediates the internalisation of CD44 and its subsequent degradation especially at hyperosmolality. Thus, it is hypothesised that *cgr*C10orf93-like/*cgr*Ttc40-like is a possible interacting protein between LPR1 and CD44 that would led to a decreased overall CD44 presentation and parallel cell-cell detachment. This would cover the observations, which were made comparing 25-CHO-S/2.5C8 as well as CHO-S-E400/1.104 on their *cgr*C10orf93-like/*cgr*Ttc40-like expression level (Figure 4.19). The relative as well as significant increase of the expression level of *cgr*C10orf93-like/*cgr*Ttc40-like in CHO-S-E400/1.104 cells at early stationary phase and the unchanged aggregation behaviour of CHO-S-E400/1.104 over the whole culture period leads to the suggestion that *cgr*C10orf93-like/*cgr*Ttc40-like could crucially (down-)regulate transmembrane receptors such as CD44. Furthermore, LRP1 was shown to mediate the maturation of the focal adhesion receptor subunit β1-integrin, whereby LPR1-positive CHO-K1 cells revealed higher cell-matrix adhesion capacity than negative CHO-K1 cell lines (Salicioni 2003). Therefore, an additional step towards LPR1-downregulation might be observed in *cgr*C10orf93-like/*cgr*Ttc40-like-overexpressing CHO-K1-derived cell lines inhibiting cell-matrix attachments by integrin complex-formation.

Taken together, it is proposed that *cgr*C10orf93-like/*cgr*Ttc40-like was partly quenched or masked once overexpressed in the highly aggregating K20-3 producer cell line, whilst in non-aggregating 25-CHO-S/2.5C8 cell line overexpression would be more toxic due to higher availability of unbound *cgr*C10orf93-like/*cgr*Ttc40-like. In the contrary, the discrepancy between overexpression of *cgr*Ttc36 in 25-CHO-S/2.5C8 and K20-3 cell line could be explained in the same but reversed manner (Figures 4.37 and 4.39). Here, the reduction of the growth supportive effect of *cgr*Ttc36 as well as the growth inhibitory effect of *cgr*C10orf93-like/*cgr*Ttc40-like could be explained due to appearance and the resulting initiation competition of the identical promoter for hIgG expression in K20-3. In contrast, the clonal *cgr*Ttc36-expressing K20-3 cell lines 1.7C1 and 1.17D7 revealed overlapping specific productivity characteristics (Figure 4.44) upon different *cgr*Ttc36-mRNA levels (Figure 4.46) as well as identical aggregation behaviour (Figure 4.45). Moreover, comparing the clonal cell lines without significant *cgr*Ttc36-overexpression (1.7E4 and 1.7G8), the cell specific productivities are comparable with the *cgr*Tt36-overexpressing K20-3 clones 1.7C1 and 1.17D7, but upon reduced maximal cell density as well as maintained aggregation characteristics. Regarding the K20-3 clone 1.5F5, the *cgr*Ttc36 expression was shown to be increased followed by maintenance of the aggregation as well as the rather low cell density and an increased cell specific productivity (Chapter 4.3.4.4). Hence, another intrinsic factor could have managed to quench the increased *cgr*Ttc36 levels leading to this phenotype. Therefore and in contrast to *cgr*Ttc36, *cgr*C10orf93-like/*cgr*Ttc40-like might inhibit cell growth in a dose dependent manner and depending on the cell aggregation grade of the cell line. Most interestingly and nevertheless, *cgr*C10orf93-like/*cgr*Ttc40-like as well as *cgr*Ttc36 showed the same aggregation phenotype leading to the suggestion that a common dose-dependent mechanism occurred.

Subsequently, comparing the sequences of *cgr*Ttc36 (XM_003510512.1) and *cgr*C10orf93-like/*cgr*Ttc40-like (XP_003510630.1), a common short and highly homologous sequence was found (Table 5.17). This ΨRAAEXG sequence (Ψ: hydrophobic residue) is located within the N-terminal TPR motif of *cgr*Ttc36 (Figure 5.2) or after the FATC domain in *cgr*C10orf93-like/*cgr*Ttc40-like (Figure 5.13), whilst both regions were predicted to be α-helical structures.

Table 5.17: Sequence alignment of *cgr*Ttc36 (XM_003510512.1) and *cgr*C10orf93-like/*cgr*Ttc40-like (*cgr*Ttc40l; XP_003510630.1). Homologous as well as functionally similar residues are shaded in red and grey, respectively.

	Domain	Alignment
cgrTtc36	TPRα	48 SKALELQGVRAAEAG-DLHTALEKF 71
cgrTtc40l	-	756 SQYAMGNWLRAAEIGQDLGESWVVQ 780

This in common, the sequences were found to possess similarities to the tyrosine protein kinases specific active-site signature of c-Src kinase (Schmidt-Ruppin A-2 avian sarcoma viral oncogene homolog) and its family members Yes (Yamaguchi sarcoma viral oncogene homolog), Fgr (Gardner-Rasheed feline sarcoma viral oncogene homolog), Blk (B lymphoid tyrosine kinase), Lck (lymphocyte-specific protein tyrosine kinase) and Hck (hemopoietic cell kinase) (Figure 5.14 and table 5.18). All Src non-receptor kinase family members above possess high homologues regions but the N-terminal sequences, which possibly mediate the specificity in receptor as well as non-receptor protein binding (Kefalas *et al.* 1995). The two isoforms of c-Src in CHO-K1 (*cgr*Src1/*cgr*Src2; simplified: *cgr*Src) possess the identical active-site sequence. The tyrosine kinase c-Src is a multidomain kinase and predominantly acts as a regulator of cell-matrix (focal adhesions) as well as cell-cell adhesions (e.g. adherens junctions) (Tsukita *et al.* 1991). Upon adhesion signals, c-Src is activated by translocation to focal adhesions or adherens junctions, whist in human homologues the inhibitory phosphorylation of C-terminal tyrosine 527 is either reversed by dephosphorylation or masked by adapter proteins leading to tyrosine 416 phosphorylation and activation of c-Src (Courtneidge 1985; Cooper and King 1986).

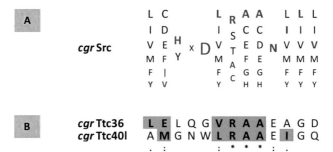

Figure 5.14: Homology of *cgr*Ttc36 and *cgr*C10orf93-like/*cgr*Ttc40-like to tyrosine protein kinases specific active-site signature of c-Src kinase (*cgr*Src). A: tyrosine protein kinases specific active-site signature consensus with highlighted sequence for *cgr*Src (XP_003497146.1 or XP_003497147.1) in red. The active-site aspartate and a downstream asparagine are mandatory. B: Tyrosine protein kinases specific active-site signature with *cgr*Ttc36 and *cgr*C10orf93-like/*cgr*Ttc40-like. Residues overlapping the active-site signature are shaded in red.

Within the inactive state of c-Src, its SH2 domain is tightly bound to the phosphorylated Y527, which can compete with phosphoproteins bearing a binding site for the N-terminal SH3 domain (PXXP). In a next step, c-Src is fully activated by phosphorylation of the tyrosine residue at 416 (Y416) (Roskoski 2004). In contrast, phosphorylation at Y416 in closed conformation represses c-Src (Irtegun *et al.* 2013). Therefore and due to the similarity to the active-site signature of *cgr*Src, it is assumed that *cgr*Ttc36 as well as *cgr*C10orf93-like/*cgr*Ttc40-like could inhibit or stabilise *cgr*Src-mediated phosphorylation in effector proteins such as kinase (e.g. focal adhesion kinase, Fak/FAK), structural proteins (paxillin, talin1) and several guanine nucleotide exchange factors (GEFs) as well as GTPase-activating proteins (GAPs).

Table 5.18: Sequence alignment of c-Src family members with *cgr*Ttc36 and *cgr*C10orf93-like/*cgr*Ttc40-like (*cgr*Ttc40l). Homologous as well as functionally similar residues are shaded in red and grey, respectively.

	Accession number	Alignment	
*cgr*Ttc36	XM_003510512.1	46	EQSKALELQGVRAAEA--G-DL 64
*cgr*Ttc40l	XP_003510630.1	754	SLSQYAMGNWLRAAEI--GQDL 773
*cgr*Src1	XP_003497146.1	380	VERMNYVHRDLRAANILVGENL 401
*cgr*Src2	XP_003497147.1	386	VERMNYVHRDLRAANILVGENL 407
*cgr*Yes	XP_003513692.1	385	IERMNYIHRDLRAANILVGENL 406
*cgr*Fgr	XP_003511485.1	363	MERMNYIHRDLRAANILVGEQL 384
*cgr*Blk	XP_003498022.1	345	LERMNSIHRDLRAANILVSETL 366
*cgr*Lck	XP_003500703.1	422	IEEQNYIHRDLRAANILVSDTL 443
*cgr*Hck	XP_003502348.1	429	IEQRNYIHRDLRAANILVSASL 450
			. :***: . *

Besides the regulation of cell-matrix adhesion as well as motility, c-Src regulates e.g. cell cycle, metabolism, apoptosis and cell-cell adhesion (Ferrando *et al.* 2012). Among these adhesion-linked proteins, distinct sequence homology to the tyrosine protein kinase-specific active-site signature was found (Table 5.19), which emphasises the putatively regulatory role of this ΨRAAEXG motif. Especially, the *cgr*Ttc36 core motif (RAAEA) was found in several cell-matrix as well as cell-cell adhesion-associated proteins such as *cgr*RhoGef2l (rho guanine nucleotide exchange factor 2-like), *cgr*Anks6 (ankyrin repeat and SAM domain-containing protein 6), *cgr*RasGap1 as well as *cgr*Rasgap4 (ras GTPase-activating protein 1/4).

Downregulation of the human homologue of CHO-K1 guanine nucleotide exchange factor (GEF) *cgr*RhoGef2l, ARHGEF2 (GEF-H1), was accompanied with upregulation of adherens junction proteins E-cadherin as well as α/γ-catenins (Cheng *et al.* 2012). Upregulation of ARHGEF2/GEF-H1 (Rho/Rac guanine nucleotide exchange factor 2) would thus result in dowregulation of adherence junction molecules and cell-cell detachment in suspension cell lines.

Table 5.19: Sequence alignment of putative (Ψ)RAAEXG[†] motif-bearing proteins with *cgr*Ttc36 and *cgr*C10orf93-like/*cgr*Ttc40-like (*cgr*Ttc40l). Homologous as well as functionally similar residues are shaded in red and grey, respectively.

	Accession number	Alignment	
*cgr***Ttc36**	XM_003510512.1	51	LELQGVRAAEAGDL	64
*cgr***RhoGef2l**	ERE87437.1	818	RRLAEERAAEASNL	835
*cgr***Anks6**	ERE83336.1	42	AGPEAARAAEAGAP	59
*cgr***RasGap1**	ERE85003.1	1	-----MMAAEAGSE	9
*cgr***RasGap4**	EGW01842.1	959	IKLASIRAAEKVEE	972
*cgr***Ttc40l**	XP_003510630.1	759	AMGNWLRAAEIGQDL	773
*cgr***Ankrd5**	XP_003495313.1	61	GCTPTMRAAELGHEL	75
*cgr***Man2**	XP_003499415.1	519	IMESRLRAAEILYYL	533

† Ψ: hydrophobic residue

RhoGef2l: rho guanine nucleotide exchange factor 2-like; Anks6: ankyrin repeat and SAM domain-containing protein 6; RasGap: ras GTPase-activating protein; Ankrd5: ankyrin repeat domain-containing protein 5; Man2: alpha-mannosidase 2

Nevertheless, the homologous proteins *hsa*ARHGEF2/GEF-H1 (AAH20567.1) (RRLGEERATEAGSL), *rno*RhoGef2 (NP_001012079.1) (QRLGEERATEAGSL) and *mmu*RhoGef2 (Q60875.4) (QRLGEERATEAGSL) bear a different motif than observed in *cgr*RhoGef2l (Table 5.19). Hence, this specific motif is poorly conserved in RhoGef2 homologues and possible futile in function. Regarding *cgr*Anks6, the motif sequence is even less conserved in human (*hsa*ANKS6, NP_775822.3) (AGPGAAAAGAVGAP) and rat homologue (*rno*Anks6, NP_001015028.2) (AGPEAARAVEAGTP), but more homologous to *cgr*Ttc36 in mouse (*mmu*Anks6, NP_001019307.1) (AGPEAVRAAEAGAP). Thus, in mouse as well as hamster, Anks6 could perform a putatively similar role in the hypothesised c-Src substrate binding that overexpression showed to induce polycystic kidney disease at least in rats (Neudecker *et al.*

2010). Primarily through its ankyrin repeats that do not compromise the (V)RAAEAG motif, ANKS6 was shown to interact with the protein kinase NEK8/NPHP9 (Hoff *et al.* 2013) in HEK293T cells, whose downregulation vulnerables mice for polycystic kidney disease and leads to reduced E-cadherin expression (Natoli *et al.* 2008). Therefore, *cgr*Anks6 activation as well as upregulation could induce cell detachment. The role of the common (V)RAAEAG motif between *cgr*Anks6/*mmu*Anks6 as well as *cgr*Ttc36 needs surely more investigation, since no direct linkage between Nek8/NEK8 and Anks6/ANKS6 with c-Src could be identified.

RasGap's (regarding *cgr*RasGap1 and *cgr*RasGap4) are GTPase activating proteins and interactors of ras proteins, which are widely considered in regulation of focal adhesion formation (Sharma 1998; Hecker *et al.* 2004). RasGAP1 (RasGAP^{p120}) negatively regulates c-Ras and for its part is regulated by phosphorylation through focal adhesion kinase (FAK), which is activated by c-Src phosphorylation (Ferrando *et al.* 2012). In addition, RasGAP1 was observed to interact with Src-phosphorylated RhoGAP^{p190} as well as FAK (Sharma 1998; Roof *et al.* 1998; Hu and Settleman 1997; Endo and Yamashita 2009) and results in RasGap1 inactivation (Moran *et al.* 1991). Therefore, it is proposed that *cgr*RasGap1 could inhibit the Src-mediated phosphorylation of *cgr*RhoGap^{p190} in a limited manner. Moreover, c-Src phosphorylates the Rap1 guanine nucleotide exchange factor (RapGEF) C3G and mediates Rap1 activation as well as adherens junction formation (Birukova *et al.* 2013; Pannekoek *et al.* 2009). Therefore and in comparison to *cgr*Ttc36, it is proposed that *cgr*RasGap1 could inhibit the Src-mediated phosphorylation of *cgr*RhoGap^{p190} and hamster RapGefs in a limited as well as dose-dependent manner by its N-terminal MMAAEAG motif. Interestingly, the *hsa*TTC36 sequence does not possess the VRAAEAG motif, but VMAAEAG, which is even more similar to the motif of *cgr*RasGap1 and *hsa*RasGAP1. Regarding *cgr*RasGap4 no relevant publications and binding partners could be revealed to date, but a similar function to *cgr*RasGap1 is suggested.

For *cgr*C10orf93-like/*cgr*Ttc40-like, a highly homologous sequence was found in *cgr*Ankrd5 (ankyrin repeat domain-containing protein 5) (Table 5.19). The function of Ankrd5/ANKRD5 could not be revealed, but the sequence MRAAELGHEL is conserved in *mmu*Ankrd5 (NP_783598.1), *rno*Ankrd5 (NP_001099986.1) and *hsa*ANKRD5 (NP_071379.3), which is located within the first ankyrin repeat region similarly to the previously described *cgr*Anks6. Ankyrin repeats were found in proteins associated with adherens junctions such as Ilk/ILK (integrin-linked protein kinase) (Vespa *et al.* 2005). Using the Conserved Domain Database (CDD) (Marchler-Bauer *et al.* 2013), *cgr*Ankrd5, *cgr*Anks6 as well as *cgr*Ilk (NP_001233743.1) were predicted to possess a common motif ([H/Ψ]XAAXXGΠ; Ψ/Π: hydrophobic/hydrophilic residue) within their first ankyrin repeat motif. Interestingly, this N-terminal region was shown to crucially mediate localisation/binding of ILK to adherens junctions (Vespa *et al.* 2005). Deletion of this the first ankyrin repeat motif and especially the part bearing the revealed common motif led to improper localisation at cell-cell junctions. In combination to all other suggestions and findings, it is vaguely hypothesised that *cgr*C10orf93-like/*cgr*Ttc40-like as well as *cgr*Ttc36 thus could mediate the dissociation of ankyrin repeat-containing proteins, similarily to results using peptidic inhibitors (NSGNGAVEDRKPSGL) mimicking the CD44 cytoplasmic ankyrin-binding domain (Zhu and Bourguignon 1998).

The N-glycosylation pathway protein cgrMan2 (α-mannosidase 2) or its binding partners could eventually be targets for cgrC10orf93-like/cgrTtc40-like. The inhibition of α-mannosidase 2 was reported to promote E-cadherin-mediated cell-cell adhesion in HCT-116 colon cancer cells (Freitas Junior *et al.* 2011). Moreover, the same observation was made in suspension Jurkat cells but using siRNA-mediated knock-down of the homologous protein Man2c1 (Qu *et al.* 2006). The comparisons with Jurkat T cells are usually more favourable because of their permanent growth in suspension and rather low aggregation behaviour. Therefore, the putative stimulation of cgrMan2 would lead to an altered more favourable product quality in biopharmaceutical protein production towards complex-type glycans avoiding accumulation of mannose-rich glycans (e.g. $Man_9GlcNAc_2$ or $Man_5GlcNAc_2$) (Yu *et al.* 2011; Lewis *et al.* 2013, supplementary figure S3). Hypothetically, cgrC10orf93-like/cgrTtc40-like and putatively cgrTtc36 as well could lead to an improved product quality mediated by cgrMan2 activation/upregulation.

In conclusion, cgrTtc36 and cgrC10orf93-like/cgrTtc40-like could be possibly quenched by adherens junction as well as focal adhesion proteins. The expression levels of both mRNAs during overexpression in 25-CHO-S/2.5C8 as well as K20-3 were comparable (data not shown), suggesting binding of both proteins to cell membrane mediating the inhibition of cell-cell adhesion. In parallel, the residual "soluble" protein could mediate to growth promoting (cgrTtc36) or inhibitory effects (cgrC10orf93-like/cgrTtc40-like) within the used CHO-K1 cell lines. The molecular mechanisms on the inhibition of cell-cell adhesion as well as the resulting effects regarding e.g. product quality need to be experimentally elucidated, but the present hypothesis could give first hints for finding the needle in the CHO haystack.

5.4 Arguments on cgrGas5 and its intronic sequence cgrSnord78

In chapter 4.3.3, findings regarding the overexpression of cgrGas5 and its intronic sequence cgrSnord78 were presented. Subsequently, the observations are discussed in chapter 5.4.1 and 5.4.2.

5.4.1 Insights into the function and structure of cgrGas5

The long non-coding RNA (lncRNA) Gas5/GAS5 consists of several intronic sequences bearing small nucleolar C/D box RNAs (C/D box snoRNA; Snord/SNORD) (Figure 5.15). Especially, the Gas5/GAS5 species of *H. sapiens* (*hsa*GAS5) and *M. musculus* (*mmu*Gas5) are well documented. In both cases, this lncRNA promotes growth arrest by inducible expression upon physical as well as chemical stress, high cell density and nutrient deprivation (Qiao *et al.* 2013; Smith and Steitz 1998). Hence, the downregulation of *hsa*GAS5 is associated with stress- as well as multidrug-resistant cancer formation and promotes tumour cell growth, whilst overexpression of *hsa*GAS5 was shown to inhibit tumour progression (Lu *et al.* 2013; Liu *et al.* 2013c; Shi *et al.* 2013; Özgür *et al.* 2013). Depending on the source (NCBI/Smith and Steitz 1998), the murine Gas5 consists of four/twelve exons and three/eleven introns, which bear in total nine C/D box snoRNAs (*mmu*Snord74, 75, 76, 44, 78, 80 and 81). In parallel, the exon/intron structure, size as well as snoRNA-content in *rno*Gas5 is widely similar to *mmu*Gas5 (Figure 5.15) and putatively in function as well even no published data is available for this species. The pre-lncRNA of human homologue *hsa*GAS5 gene includes an additional C/D box snoRNA (*hsa*SNORD77) within the fourth intron and is overall the longest Gas5/GAS5 species (Smith and Steitz 1998; Figure 5.15).

Compared to *hsa*GAS5, *mmu*Gas5 and *rno*Gas5, the overall size of full-length *cgr*Gas5 (pre-lncRNA) is massively truncated possibly during reiterated culture *in vitro*, whose changes could be caused by chromosomal changes along with somatic evolution as well as extent subculture and subcloning of CHO-K1 (Kao *et al.* 1969). Comparing other C/D box snoRNA-bearing lncRNA such as the U87HG, SNHG1 and RPS8 genes (Makarova and Kramerov 2009), evolutionary changes in size and content are massive to some extent (U87HG and SNHG1, but not RPS8), but within placental vertebrates and especially rodents the changes are marginal. As well, the intrinsic copies of snoRNA species were evolutionary reduced towards higher placental vertebrates (Makarova and Kramerov 2009). For instance, the *D. rerio* (zebrafish, *dre*) gas5 transcript derived from genomic DNA (NW_001879345.1) possesses each two juxtaposing copies of *dre*Snord75, *dre*Snord79, *dre*Snord80 and *dre*Snord47 next to single copies of *dre*Snord74, *dre*Snord76, *dre*Snord44 as well as *dre*Snord78 (Higa-Nakamine *et al.* 2011). Here, the intron alignment is different to e.g. *hsa*GAS5 as well.

Nevertheless, among human, mouse, fugu and zebrafish Gas5/GAS5 the similarities in size and snoRNA composition are apparent (Higa-Nakamine *et al.* 2011, supplementary figure S1). Most interestingly, the *C. griseus* strain 17A/GY chromosome 5 chr5_contig_16985 (gi|530146579; gb|APMK01011234.1|200-4400) as well as the contig061522 of its proximal relative *Mesocricetulus auratus* (*M.auratus*) (gi|471627016; gb|APMT01061522.1|9200-13000) revealed very homologous sequences and similar structures to *mmu*Gas5 and *rno*Gas5. The *C. griseus* 17A/GY and *M.auratus* genomic sequences possess all C/D box snoRNAs observed in *mmu*Gas5 and *rno*Gas5, whilst Snord77 is absent again (alignment not shown). In addition, connecting the CHO-K1- (ATCC/ECACC), CHO-DG44- as well as CHO-S-derived *C. griseus* scaffold1279_36 (gi|521843635; gb|AMDS01079492.1) and scaffold1279_37 (gi|521843634; gb|AMDS01079493.1) (entries from 09. July 2013, Lewis *et al.* 2013) revealed the same alignment on their negative strands, proving that the previous genomic data was incomplete and resulted in misinterpretation. Therefore, the truncation was experimentally performed on basis of *C. griseus* C41583244_1 whole genome shotgun sequence (gi|342371527; gb|AFTD01262011.1, from 03. August 2011). Nevertheless, the amplification of *cgr*Gas5$_{2029bp}$ and its splice variants was performed in all conscience and state of knowledge of that period. Furthermore, the P$_{CMV}$-mediated overexpression full-length *cgr*Gas5 (*cgr*Gas5$_{2029bp}$) did not lead to a significant growth arrest (Figure 4.28), in contrast to findings regarding *hsa*GAS5 and *mmu*Gas5. Moreover, downregulation of *hsa*GAS5 improved culture performance of CEM-C7 T-leukaemic cells (Williams *et al.* 2011).

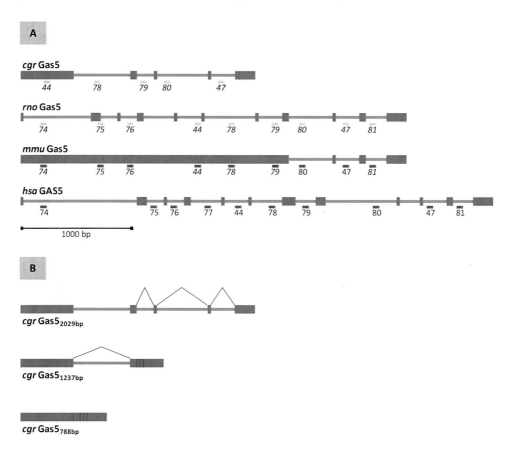

Figure 5.15: Schematic overview of different Gas5/GAS5 species. A: The full-length Gas5/GAS5 species of *C. griseus* (*cgr*Gas5; 2029 bp; gi|342371527), *R. norvegicus* (*rno*Gas5; 3322 bp; gi|389669371), *M. musculus* (*mmu*Gas5; 3327 bp; gi|34740397), *H. sapiens* (*hsa*GAS5; 4087 bp; gi|29498337) as well as their exonic/intronic regions (grey/red parts) are shown in a scaled manner. Small nucleolar C/D box RNA species (Snord/SNORD) within the introns or exons are either depicted in bars and the respective number (e.g. 44: Snord44 or SNORD44), highlighting the grade of documentation (Dark grey: listed on NCBI or published; Light grey: predicted homologues). Introns were revealed by alignment of genomic sequences with documented (*hsa*GAS5: NR_002578.2; *mmu*Gas5: NR_002840.2; *rno*Gas5: NR_002704.1) or experimental (*cgr*Gas5) splice variants. In earlier publications (Smith and Steitz 1998), *mmu*Gas5 was shown to be fully spliced analogous to *rno*Gas5. Here, the official splice variant of *mmu*Gas5 is displayed. B: Schematic view of full-length *cgr*Gas5 (*cgr*Gas5$_{2029bp}$) and its two experimentally proven splice variants (*cgr*Gas5$_{1237bp}$ and *cgr*Gas5$_{788bp}$). Further details are shown in table 4.2.

Compared to the control cell line, 25-CHO-S/2.5C8/*cgr*Gas5$_{2029bp}$ and all other *cgr*Gas5 variant-overexpressing cell lines revealed higher viabilities suggesting the absence of the cell inhibitory sequence or snoRNA. Since C/D box snoRNA species *cgr*Snord74, *cgr*Snord75, *cgr*Snord76, *cgr*Snord77 as well as *cgr*Snord81 are absent in *cgr*Gas5$_{2029bp}$ and the exonic sequences even between *mmu*Gas5 as well as *rno*Gas5 are poorly conserved, the absent snoRNAs are shortlisted. The C/D box snoRNA *cgr*Snord44 might be supposed to cause pro-apoptotic and growth-arresting signals as well, because of the absent C/D box snoRNAs by truncation and the putative incapability of splicing. Snord44/SNORD44 is widely considered as transcription analysis control for small RNA species and its transcription is rather stable in probes of healthy as well as cancer patients (Appaiah *et al.* 2011). Nevertheless, splice inhibition as well as antisense quenching led to massive phenotypic changes in embryonal *D. rerio* development (Higa-Nakamine *et al.* 2011). Therefore, Snord44/SNORD44 probably occupies an important position in complex organisms including apoptosis, which crucially mediates development as well as embryogenesis (Abrams *et al.* 1993). As the sole C/D box snoRNA in Gas5/GAS5, Snord44/SNORD44 mediates the 2'-*O*-methylation at A166 (*mmu*) or A170 (*hsa*) of pre-18S ribosomal RNA (rRNA) in a site-specific manner (Kiss-László *et al.* 1996; Smith and Steitz 1998), whose inhibition by downregulation putatively leads to impaired ribosome subunit maturation as well as export (White *et al.* 2008; Liang *et al.* 2009). On the other hand, *cgr*Snord44 is suggested to be intrinsically unsplicable and inactive therefore. Thus, the overexpression of *cgr*Snord44 would not improve the growth of the cell neither.

Besides the absent or putatively inactive C/D box snoRNAs, *cgr*Snord78, *cgr*Snord79, *cgr*Snord47 as well as *cgr*Snord80 were correctly spliced and possibly active therefore. Comparing the splice variants *cgr*Gas5$_{2029bp}$, *cgr*Gas5$_{1237bp}$ and *cgr*Gas5$_{788bp}$, the overexpression of present full-length *cgr*Gas5 (*cgr*Gas5$_{2029bp}$) showed a similar growth characteristic like the fully spliced ncRNA (*cgr*Gas5$_{788bp}$), but slightly differing in viability (Figure 4.28). Nevertheless, both variants seemed to be slightly inhibitory compared to the control suggesting that the splicable C/D box snoRNAs are functionally levelling each other. Moreover, the *cgr*Snord78-bearing *cgr*Gas5$_{1237bp}$ would putatively implicate that *cgr*Snord78 would quench the inhibitory effects of *cgr*Snord79, *cgr*Snord47 and *cgr*Snord80. Anyhow, among these putatively inhibitory C/D box snoRNAs other beneficial functions could be revealed by separated transcription or direct introduction of each oligonucleotide. Notably to mention, the nutrient- as well as growth phase-independent *cgr*Gas5 transcription in CHO-S-E400/1.104 and 25-CHO-S/2.5C8 unveiled a non-inducible manner of regulation in CHO-K1 (Figure 4.19), which was emphasised by unaltered Ct-values determined in mutant as well as control cell line (data not shown). Therefore, this observation strongly leads to the assumption that *C. griseus* Gas5 is possibly poorly regulated by partial nutrient deprivation or cell density stress (Smith and Steitz 1998). Nonetheless, the data was monitored until the early stationary phase (Figure 4.16) and *cgr*Gas5 could be differentially expressed in late stationary as well as death phase of a culture. Due to the putatively tight connection of fully spliced *cgr*Gas5 ncRNA with nonsense-mediated decay mechanisms, the expression of the involved UPF1 ATPase *C. griseus* homologue need further be evaluated (Mourtada-Maarabouni and Williams 2013).

Thus, the putatively fully spliced *cgr*Gas5 ncRNA might mediated the growth arrest in *H. sapiens* and *M. musculus*, but a possibly higher amount of *cgr*Upf1 in CHO-K1 would decrease the inhibitory effect. Furthermore, the unspliced *cgr*Snord47-exon5 fragment of *cgr*Gas5 was monitored in all cases using the respective qRT-PCR primer pair (Table 3.8). Hence and due to the undetectable *cgr*Snord47-unspliced *cgr*Gas5 variant within total RNA pool at early stationary phase of CHO-S-E400/1.104, which represented the basis of preparative cDNA, the different variants could be differentially transcribed. Moreover, the unmonitored isolated intronic *cgr*Snord-species would possess to most interesting transcription differences, since the fully spliced *cgr*Gas5 variant (putatively *cgr*Gas5$_{788bp}$) might be rapidly degraded and downregulation by nonsense-mediated decay (Tani *et al.* 2013). By this mechanism, the snoRNA levels are assumed to be enriched by trend. Therefore, more effort is aimed regarding the determination of transcription differences of the *cgr*Gas5-derived C/D box snoRNAs.

Looking forward, the present work revealed some first insights in Gas5 function within the CHO-K1-derived cell lines. Further studies are required to fulfil the complexity of the multiple snoRNA-bearing lncRNA *cgr*Gas5 mechanisms. The putative function and molecular mechanism of a single *cgr*Gas5-derived C/D box snoRNA (*cgr*Snord78) should be further unveiled in chapter 5.4.2.

5.4.2 Possible function and targets of *cgr*Snord78

The function of Snord78/SNORD78 is hardly documented. Recent results in *D. rerio* embryos revealed a slight effect on embryogenesis upon inactivation of *D. rerio* C/D box snoRNA 78 (*dre*Snord78) mediated by inhibition of splicing (Higa-Nakamine *et al.* 2011). This splicing inhibition of *dre*Snord78 intronic sequence in *D. rerio* embryos revealed a slight phenotypic change compared to the control (Higa-Nakamine *et al.* 2011, supplementary figure S5). Here, the oligonucleotidic splice inhibitor of Snord78 intron affected the head region of the *D. rerio* embryo shown by a malformed 4th ventricle. Nevertheless, the overall effect was rather low and the documented embryos upon control as well as antisense oligonucleotide treatment showed a similar systemic alteration in their yolk sacs namely an oedemic deposit (Higa-Nakamine *et al.* 2011, supplementary figures S4 and S5). Hence, the overall vitality of these *D. rerio* embryos might be possibly perturbed leading to misinterpreted conclusions (personal communication). In addition, the GAS5-associated snoRNAs *hsa*SNORD44, *hsa*SNORD76 and especially *hsa*SNORD78 were upregulated upon virus infection suggesting a supportive effect on viral transcript translation due to the resulting increased demand on translational capacity as well as efficiency (Saxena *et al.* 2013). This assumption would strengthen the present results of *cgr*Snord78 on cell specific productivity enhancement in CHO-K1-derived cell lines (Figure 4.31). Recently, the H/ACA box snoRNA U19 (*cgr*Snora19) was found to slightly induce the pseudouridylation of 28S ribosomal RNA (rRNA) leading to increased ribosome efficiency and translation activity as well as increase the overall polysome content once overexpressed in CHO-DG44 (Courtes *et al.* 2014). As a result and in parallel to the observations, which were made by *cgr*Snord78 overexpression, the cell specific productivity was increased with simultaneous maintenance of the growth rate. The intronic H/ACA snoRNA *cgr*Snora19 (U19) was documented to essentially mediate the pseudouridylation of 28S rRNA at two distinct sites (Bortolin and Kiss 1998). In contrast to H/ACA snoRNA, the C/D box snoRNAs are known to mediate ribosome maturation by binding of ribosome subunit and following 2'-*O*-methylation at the fifth nucleotide upstream to the D/D' box (CUGA) (Table 4.4; Figure 5.16) (Kiss-László *et al.* 1996).

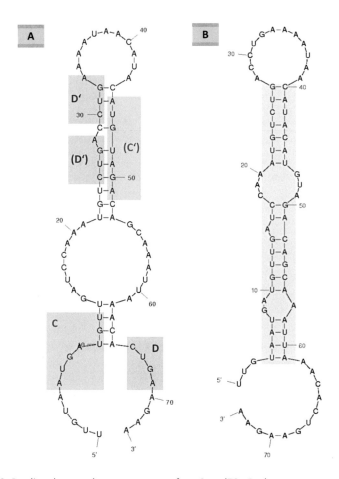

Figure 5.16: Predicted secondary structures of *cgr*Snord78. Both structures were predicted and plotted by Mfold (Zuker 2003) (Structure A: dG = -7.30; Structure B: dG = -8.00). A: The putative C (AUGAUGU), C' (AUGUAGA) as well as the D'/D consensus sequences (CUGA) are shaded in red. Undocumented but possible C/D box sequences are shown in brackets. A putatively additional D' sequence could vary the 2'-*O*-methylation directed by residue 20. The other documented D'/D sequences (Smith and Steitz 1998) would mediate a possible ribose 2'-*O*-methylation of complementary sequences directed by residues 25 and 61, respectively. All C/D'/C'/D sequences were predicted regarding published consensus sequence motifs (van Nues *et al.* 2011). B: This energetically more probable structure possibly habours a putative miRNA-stem-loop structure (shaded in grey).

Despite the mechanistic differences, the observed effects in overexpression of *cgr*Snora19 and *cgr*Snord78 were similar leading to the assumption that in Intron*cgrSnord78*-overexpressing CHO-K1 cell lines the ribosome efficiency is increased comparable to the reported *cgr*Snora19-overexpressing CHO-DG44 cells. Thus, the ribosome targets need to be determined further estimating the putative mechanism of *cgr*Snord78. The C/D box snoRNA *hsa*SNORD78 (U78) was reported to bind the human 28S rRNA subunit and 2'-*O*-methylate ribosome residue at position 4583 (Smith and Steitz 1998), but no published implication upon this post-transcriptional modification could be found. The human *hsa*SNORD78 (NR_003944.1) strongly differ from *cgr*Snord78 assuming another function of this snoRNA species in *C. griseus* as well as CHO-K1 (Table 4.4). Thus, alignment of the sequences upstream D'/D of *cgr*Snord78 with the *C. griseus* ribosomal RNA subunits 28S (*cgr*28SrRNA; gi|349585429|NR_045212.1), 18S (*cgr*18SrRNA; gi|346716260|NR_045132.1) and 5.8S (*cgr*5.8SrRNA; gi|346716261|NR_045133.1) revealed a putative binding of *cgr*18SrRNA as well as *cgr*18SrRNA (Figure 5.17).

Figure 5.17:   Putative ribosomal RNA (rRNA) targets of *cgr*Snord78. Alignments of *C. griseus* ribosomal RNA subunits 28S (*cgr*28SrRNA; gi|349585429|NR_045212.1), 18S (*cgr*18SrRNA; gi|346716260|NR_045132.1) and 5.8S (*cgr*5.8SrRNA; gi|346716261|NR_045133.1) (shaded in red) with *cgr*Snord78 reveals *cgr*28SrRNA and *cgr*18SrRNA as putative targets for 2'-*O*-methylation. The putative modified residue of *cgr*28SrRNA and *cgr*18SrRNA as well as the C/D'/C'/D sequences are highlighted. Alignments with the highest alternating coverage are shown only.

The *cgr*Snord78-mediated 2'-*O*-methylation was predicted to be rather unlikely due to a very short complementary sequence, which was under the known limit (ten nucleotides) and too far away from the D' boxes. Nevertheless, *cgr*Snord78 could mediate the folding as well as recruitment of *cgr*5.8SrRNA as well as leading to a facilitated assembly of the ribosomal complex (Dutca *et al.* 2011). By these theoretical findings, the overexpression of *cgr*Snord78 would putatively support as well as enhance the 2'-*O*-methylation of *C. griseus* 28S and 18S ribosomal RNA and thus increase the pool of active ribosomes leading to improved translation capacity. Alternatively, by binding of *cgr*Snord78 to ribosomal RNA, the assembly and folding could be improved.

Next to modifications on ribosome RNA by snoRNAs, small nuclear RNA (snRNA) splice factors are pseudouridylated as well as 2'-*O*-methylated in parallel with particularly massive implications in splice efficiency (Zhao *et al.* 2002). The splicing efficiency or the simple presence of an intron was shown to improve recombinant protein production in vitro and in vivo (Khamlichi *et al.* 1994; Whitelaw *et al.* 1991). Therefore, perturbed as well as improved intron splicing would strongly influence the titer and cell specific productivity. In the following, various snRNA splice factors, which are post-transcriptionally modified by pseudouridylation as well as 2'-*O*-methylation (Deryusheva *et al.* 2012; Yu *et al.* 1998; Dönmez *et al.* 2004), were aligned to the antisense sequence in vicinity to the D'/D box sequences. Due to the absence of *C. griseus* sequences, human homologues were aligned assuming rather conserved sequences of these crucial factors among mammals. Hence, the branch point recognising snRNAs *hsa*U2.1 (RNU2-1; NR_002716.3), *hsa*U2.2 (RNU2-2P; NR_002761.3), *hsa*U2.3 (RNU2-3P; NG_001258.2) as well as *hsa*U2.4 (RNU2-4P; NG_001257.3) and the spliceosomal-executing *hsa*U4A (V00592.1) as well as *hsa*U6 (RNU6-1; NR_004394.1) were aligned in prior to *C. griseus* genome alignment (data not shown) (Singh and Cooper 2012).

As a result, the major U2 snRNA *hsa*U2.1 possessed complementary sequences to C'/D intermediate part of *cgr*Snord78. Subsequently, *hsa*U2.1 was aligned to *C. griseus* scaffold23027 (gi|351510707|NW_003620837) leading to the homologues hamster snRNA *cgr*U2.1 (position 9152 – 9393). The latter was finally predicted to be putatively 2'-*O*-methylated at position 55 (U55) (Figure 5.18). This predicted modification, located within a stem-loop structure in vicinity and downstream of the branch point recognition sequence (Figure 5.18), is undocumented. Known proximate 2'-*O*-methylation sites in *hsa*U2.1 were documented to be at positions U47 and C61 (Dönmez *et al.* 2004; Yu *et al.* 1998). At position U54 of *hsa*U2.1, the residue is pseudouridylated, which could be positively or negatively influenced by the predicted U55-methylation. Thus, a regulatory mechanism of *cgr*Snord78 on U2-dependent spliceosome formation and preferences in branch point recognition could be assumed. Therefore, it is hypothesised that *cgr*Snord78 could facilitate the translation efficiency by modulation of ribosomal subunit maturation as well as the splice efficiency. This latter hypothesis could be reinforced by the observation upon co-expression of intronic *cgr*Snord78 and *cgr*Ttc36 (Figure 4.49 B) by facilitating alternative splicing (Figure 5.19).

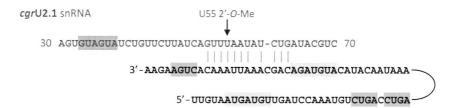

Figure 5.18: Putative *C. griseus* U2.1 (*cgr*U2.1) small nuclear RNA (snRNA) binding as well as modification mediated by *cgr*Snord78. The hamster sequence of the splice as well as branch point recognition factor *cgr*U2.1 (*cgr*U2.1snRNA) was determined by alignment of human homologue *hsa*U2.1 (gi|225735587|NR_002716.3) and *C. griseus* whole genome shotgun scaffold23027 (gi|351510707|NW_003620837|9152-9393). An arrow denotes the putative 2'-*O*-methylation site (U55 ribose). The proximate sequence involved in branch site recognition is shaded in grey. The C/D'/C'/D sequences of *cgr*Snord78 are highlighted shaded in red.

Especially, the alternative branch points within the 3'-end of IRES$_{FMDV}$ possess a higher coverage to the complementary *cgr*U2.1 branch point-binding site (Figure 5.19 B) suggesting a more beneficial splice moment upon *cgr*U2.1 stimulation by *cgr*Snord78-mediated U55-2'-*O*-methylation. The well-balanced complementary branch site residues could lead to a stronger binding of *cgr*U2.1 of this IRES$_{FMDV}$-splice site and beneficially induce spliceosomal complex formation at this point. Moreover and independently to *cgr*U2.1, the splicing of Intron$_{cgrSnord78}$ could be inhibited by NFκB1/RelA (p50/p65) homo-/heterodimer binding (Chapter 5.3.2.4) due to the predicted binding consensus sequence in vicinity of the 3'-splice site (Figure 5.19). Consequently, the coding sequence of *cgr*Ttc36 would be disrupted in each case. This mechanism could not have been seen using the utilised qRT-PCR primer pair for *cgr*Ttc36 (Table 3.8). Nevertheless, this mechanism needs further elucidation regarding splice behaviour as well as *cgr*Ttc36 protein level. For *hsa*SNORD78 and bovine Snord78, miRNA-like stem-loop structures were proposed (Guduric-Fuchs *et al.* 2012). Hence, *cgr*Snord78 could harbour miRNA-like functions as well; even the predicted secondary structure could be revealed rather short and intermitted complementary stem structures (Figure 5.16). Interestingly, *cgr*Snord78 was found to possess high coverage to *cgr*-pre-miR-10a and especially the mature miRNA *cgr*-miR-10a-5p (Figure 5.20 A), which was shown to act as a translational activator of various ribosomal protein transcripts by binding at their 5'-UTR region (Ørom *et al.* 2008). Analogously to miR-10a-5p, various ribosomal proteins could be increasingly translated upon dicer-mediated maturation of the predicted *cgr*Snord78-miRNA (Lee *et al.* 2002). Thus in combination to the 2'-*O*-methylation mechanism, a matured *cgr*Snord78-miRNA could enhance the overall translational efficiency; especially regarding possible bottlenecks such as the P$_{CMVenhanced}$-driven constitutive expression of hIgG by K20-3 increasing the cell specific productivity (Figure 4.31).

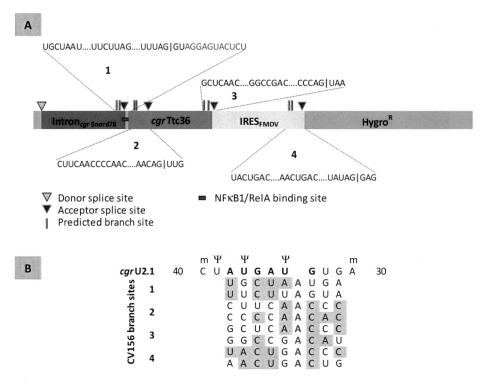

Figure 5.19: Schematic overview of possible splice sites as well as branch points in CV156-derived mRNA. A: Alternative splice sites are situated within the coding sequence of *cgr*Ttc36 or within the IRES$_{FMDV}$ (Table 3.20) resulting *cgr*Ttc36 non-sense splicing. The putative NFκB1/RelA (p50/p65) binding sequence (AGGAGUACUCU) was predicted using consensus sequence optimisation (Wong *et al.* 2011). The respective motifs (letters) and splice sites (bar) are highlighted in red. B: Alignment of snRNA *cgr*U2.1 with CV156-branch sites. Complementary residues are shaded in grey. The branch points are indicated as bold red letters. Possible modifications are pseudouridylation (Ψ) and 2'-*O*-methylation (m) (after Yu *et al.* 1998).

Furthermore, the overexpression of intronic *cgr*Snord78 could perturb the activity of *cgr*-miR-10a-5p by an interceptional sponge-like mechanism (alignment not shown). Thus, knockdown of *hsa*-miR-10a was shown to downregulate IκBα in human aortic endothelial cells resulting in an increased activation as well as nuclear translocation of RelA (p65) (Weiss *et al.* 2009). As a consequence and in combination to the putatively increased *cgr*U2.1 activity, this mechanism could lead to alternative non-sense splicing in CV156-derived mRNA, which mediated the co-expression of *cgr*Snord78 and *cgr*Ttc36, due to the present NFκB1/RelA (p50/p65) homo-/heterodimer binding site in vicinity to the Intron$_{cgr Snord78}$-3'-splice site (Figure 5.19). Hence, it could be assumed that the combination with the proposed putative dicer-driven maturation of *cgr*Snord78 (Figure 5.20) might increase the translational efficiency and alternative splicing by equally imitating as well as inhibiting *cgr*-miR-10a-5p. The capability or incapability of *cgr*Snord78-miRNA in post-transcriptional inhibition of complementary mRNA species was investigated by BLAST search of intermediate C/D' or C'/D sequences against *C. griseus* mRNA sequences (Altschul *et al.* 1990) analogously to studies with artificial anti-HSPA8 interfering snoRNAs (Stepanov *et al.* 2013; Stepanov *et al.* 2012).

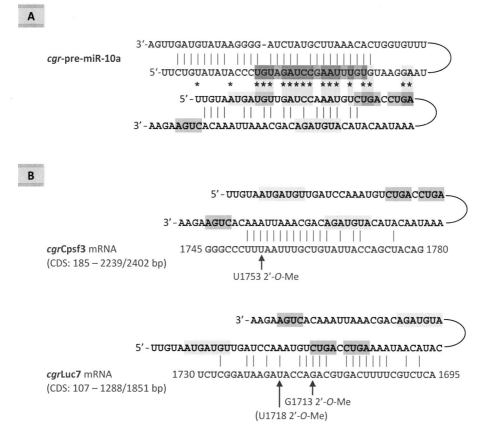

Figure 5.20: The strong coverage with *cgr*-miR-10a and putative mRNA targets of *cgr*Snord78 reveal possible mechanisms beyond ribosomal as well as small nuclear RNA regulation. A: Non-complementary alignment of *C. griseus* pre-miR-10a (*cgr*-pre-miR-10a; MI0020381) with *cgr*Snord78 reveals strong coverage between both RNA species, especially within the region of mature *cgr*-miR-10a-5p (highlighted in grey). The snoRNA *cgr*Snord78 (denoted in black letters) is shown analogously to miRNA-like structure predicted in figure 5.16 B, whereat the C/D'/C'/D sequences are highlighted in shades of red. Common residues are marked with asterisks as well as shaded in light grey. B: Putative *C. griseus* mRNA targets of *cgr*Snord78 with a role in RNA regulation with indicated complementary residues and possible 2'-*O*-methylation sites. Cpsf3 (gi|354504215|XM_003514125.1): cleavage and polyadenylation specificity factor 3; Luc7 (gi|354478788|XM_003501549.1): putative RNA-binding protein Luc7-like 1.

By BLAST search, an exceptional culmination of RNA processive as well as secretory pathway proteins such as *cgr*Cpsf3 (cleavage and polyadenylation specificity factor 3;

XM_003514125.1), cgrCpeb1 (cytoplasmic polyadenylation element-binding protein 1; XM_003512728.1), cgrMtmr6 (myotubularin-related protein 6; XM_003500514.1), cgrErgic2 (Endoplasmic reticulum-Golgi intermediate compartment protein 2; XM_003515932.1), cgrNxf2 (nuclear RNA export factor 2; XM_003507896.1) and cgrLuc7 (RNA-binding protein Luc7-like 1; XM_003501549.1) was observed. In contrary to the proposed activation of translation, the putative downregulation of each of these factors by cgrSnord78 would lead to a perturbation of post-transcriptional processes.

The cleavage and polyadenylation specificity factor 3 (Cpsf3/CPSF3) is involved in enhancement of mRNA by polyA tail extension and its downregulation was reported to inhibit the expression of several transcripts ultimately leading to cell death (Calzado et al. 2004; Zhu et al. 2009). Similarly, the cytoplasmic polyadenylation element-binding protein 1 (Cpeb1/CPEB1) enhances translation by binding to U-rich (e.g. UUUUUAU) upstream of the polyA consensus sequence (AAUAAA) and initiation of cytosolic polyadenylation as well as translation in an inducible manner (D'Ambrogio et al. 2013; Kim et al. 2011). Furthermore, the downregulation of CPEB1 was shown to simultaneously downregulate crucial adherens junction proteins such as E-cadherin and β-catenin (Chapter 5.3.4) leading to adhesion-independent growth (metastasis) (Grudzien-Nogalska et al. 2014). In pool as well as clonal K20-3/cgrSnord78 cell lines no loss in cell-cell adherence could be observed, but rather a slight increase in aggregation (Chapters 4.3.3.3 and 4.3.3.4). In addition, CPEB1 was reported to possess the capability for localisation within the nucleolar bodies and possibly mediate the ribosomal biogenesis (Ernoult-Lange et al. 2009). On the other hand, the myotubularin-related protein 6 (Mtmr6/MTMR6) and the endoplasmic reticulum-Golgi intermediate compartment protein 2 (Ergic2/ERGIC2) were demonstrated to be associated with secretory pathways as well as apoptosis (Mochizuki et al. 2013; Liu et al. 2003), but downregulation would increase cell number as well as viability. Again, these effects could not be shown in cgrSnord78-overexpressing CHO-K1 cell lines (Figures 4.29 and 4.30). The mRNA-translocation protein Nxf2/NXF2 interacts with cytosolic motor proteins associated within nuclear export and thus cytosolic localisation of mRNA (Takano et al. 2007). A putative downregulation of cgrNxf2 by the proposed cgrSnord78-miRNA would perturb the cell specific productivity, which could not be observed (Figure 4.34). And finally, the human as well as yeast homologues of U1 snRNA-associated Luc7/LUC7 act as crucial subunit within the U1-snRNP complex mediating 5'-splice site recognition (Fortes et al. 1999; Puig et al. 2007). Here, hsaLUC7 was reported to induce alternative splicing upon overexpression. Thus, downregulation of cgrLuc7 would strongly affect 5'-splice site recognition. Taken together, the miRNA-structure of cgrSnord78 would not be able to generate antisense oligonucleotides disrupting the expression of the proteins above or by another antisense mechanism neither (Stepanov et al. 2013). Moreover, the majority would rather be favourable in increasing the cell specific productivity without affecting the cell growth as observed (Figure 4.31).

For two of the presented putative targets (cgrCpsf3 and cgrLuc7), the alignment with cgrSnord78 revealed a possible 2'-O-methylation modification of their mRNA (Figure 5.20 B).

Nevertheless, studies concerning C/D box snoRNA-mediated 2'-O-methylation of branch points showed the capability to partly inactivate the target mRNA (Ge *et al.* 2010; Stepanov *et al.* 2012). Furthermore, this effect was proven to be accomplished in a D'/D box independent manner (Stepanov *et al.* 2013). Moreover, C/D box snoRNAs were shown to inhibit nucleolar but not nucleoplasmic RNA editing only, which would represent another possible mRNA regulating mechanism (Vitali *et al.* 2005). Inasmuch, polymerase II-driven mRNA transcription (e.g. by P_{CMV} and P_{SV40}) would be unaffected by C/D box snoRNA-mediated modification. Nonetheless, mRNA was shown to be translocated to nucleolar bodies upon certain stress responses (Názer *et al.* 2012), which could putatively occur downstream of native signal transduction pathways as well. Because of the observed numbered appearance of *cgr*Snord78-targeting transcripts associated in translation as well as secretion, an additional mechanism beyond the 2'-O-methylation of ribosomal and spliceosomal small nuclear RNA is very likely. This mechanism, a possible nucleolar translocation as well as RNA editing affection, could be stress-inducible as observed in *Trypanosoma cruzi* (Názer *et al.* 2012) or mediated by other yet unrevealed pathways.

The last proposed mechanism concerns the complete intronic sequence possessing *cgr*Snord78. Due to the application of the full-length *cgr*Snord78-bearing intronic sequence upon overexpression, the additional sequences flanking *cgr*Snord78 could contribute to the observed increase in cell specific productivity as well. The miRBase alignment thereof (Griffiths-Jones *et al.* 2007) revealed a putative binding site of mature miRNA *cgr*-miR-21-5p in vicinity of the 5'-splice site (Figure 5.21), which could inhibit splicing by competing spliceosomal factors. In this context, the transient overexpression of miR-21 was reported to slightly (approx. 0.8-fold) decrease the cell specific productivity in EpoFc-producing CHO dhfr⁻ DUKX-B11 cell line without affecting the growth rate or cell density (Jadhav *et al.* 2012). Thus, overexpression of the Intron$_{cgr\text{Snord78}}$ could intercept *cgr*-miR-21-5p, which could contribute to the observed increase in cell specific productivity.

Figure 5.21: Sequence alignment of *cgr*-miR-21-5p to *cgr*Gas5$_{2029bp/1237bp}$ revealing a putative binding site in *cgr*Snord78-bearing intron (Intron$_{cgr\text{Snord78}}$). The exonic as well as intronic sequences of *cgr*Gas5$_{2029bp/1237bp}$ (see also gi|342371527|gb|AFTD01262011.1) with the indicated localisation are shown in grey and red letters, respectively. The mature miRNA *cgr*-miR-21-5p (MIMAT0004417) is depicted in black letters.

In addition, *cgr*-miR-21 is upregulated in CHO-K1 in stationary phase compared to exponential phase (Gammell *et al.* 2007). A less efficient splicing of the *cgr*Snord78-bearing intron was observed in K20-3/*cgr*Snord78 pool as well as clonal cell lines at late exponential compared to early exponential phase (Figure 4.35). Thus, it could be proposed that *cgr*-miR-21 could have an influence on the splicing of Intron*cgr*Snord78 and finally inhibit the maturation of *cgr*Snord78.

In conclusion, overexpression of the natively intronic *cgr*Snord78 derived from *cgr*Gas5 significantly increased the cell specific productivity possibly by increasing the maturation of ribosomal as well as small nuclear RNAs. Additional proposed mechanisms such as translational activation of snRNPs, 2'-O-methylation of mRNA within nucleolar bodies and sponge-like interception of inhibitory miRNAs could support these functions. Overall, the overexpression of *cgr*Snord78 was revealed as a simple as well as powerful tool in improving recombinant protein production using CHO cell lines.

5.5 Optimisation and industrial relevance of *cgr*Snord78 and *cgr*Ttc36

Within the last chapter, the economic exploitability of *cgr*Ttc36 as well as *cgr*Snord78 is argued. In both cases, the overexpression of either *cgr*Ttc36 or *cgr*Snord78, the host CHO-K1 producer cell line revealed an increase in hIgG concentration either by increasing the integral viable cell density or the cell specific productivity (Chapters 4.3.3.5 and 4.3.4.5). Nevertheless, the logical combination of both factors for improving both production parameters was not beneficial and even led to a decrease in integral viable cell density in 25-CHO-S/2.5C8 (Figure 4.49). Co-overexpression *cgr*Ttc36 and *cgr*Snord78 was not performed in K20-3 producer cell line due to the inhibitory effect observed, thus productivity data are not available. Consequently, molecular as well as process-related optimisation steps are discussed for each factor separately. At the end, further suggestions on vector optimisation are focussed.

Starting with *cgr*Ttc36, the coding sequence could be optimised regarding CpG island content as well as codon optimisation, since the native coding sequence possesses clusters of several CpG islands (Figures 4.38 and 5.22). Even no CpG methylation could be observed directly, these steps could improve expression stability as well as the translational efficiency. In addition, the predicted branch points, splice sites, Cirbp as well as Rbm3 binding sites could impair the proper expression of *cgr*Ttc36 (Figure 5.11). Therefore, the sequence was manually optimised towards various binding site- and CpG island-deficiency using the proposed codon usage of CHO cells (Figure 5.22 B; Baycin-Hizal *et al.* 2012). Regarding the CpG island optimisation, possible codons of serine (TCG), proline (CCG), threonine (ACG), alanine (GCG) and arginine (CGG, CGA, CGT, CGC) need to be changed. Especially for arginine, two other codons (AGA, AGG) could be applied to fulfil the proper sequence. Therefore, the CpG islands could be efficiently abolished from optimised coding sequence without affecting the protein sequence (Figure 5.22 B). In the same manner, the overall codon usage was optimised abolishing the predicted splicing-related as well as binding sequences in parallel without affecting the protein sequence. In this order, the downstream sequences (e.g. IRES$_{FMDV}$, Figure 5.11) should be replaced or optimised regarding splice site-deficiency as well.

In addition, putative inhibitory protein sequences could be abolished by mutation (SUMOylation and phosphorylation sites), which surely needs to be experimentally elucidated first. In addition, by further elucidation of activating phosphorylations or other modifications, constitutively active variants of *cgr*Ttc36 could be engineered, since the native protein is putatively regulated by conformational changes (Chapter 5.3.2.1). Another interesting point could be the increase in affinity of *cgr*Ttc36 to *cgr*Hsp70 or even *cgr*Hsp90. Hence, the dissociation constants for each heat shock protein should be determined at first analogously to Sti1/Hop (Scheufler *et al.* 2000). Nevertheless, the ultimate mechanism behind *cgr*Ttc36-mediated effect need to be further disclosed, which was assumed to be reliant on *cgr*Hsp70 content due to the similar effect upon overexpression in CHO-K1 cell lines (Lee *et al.* 2009; chapter 5.3.2.2).

```
ATGGGGACTCCAAATGATCAGGCAGTGCTGCAGGCCATCTTCAACCCCAACACACCATTTGGAGATGTCAT
TGACTTGGACCTGGAAGAAGCAAAGAAAGAAGATGAAGATGGAGTTTTCCCTCAAGAACAGTTGGAGCAGT
CCAAAGCTCTGGAGTTGCAGGGAGTGAGGGCAGCAGAAGCTGGGGACCTCCACACAGCCCTGGAGAAGTTT
GGCCAAGCTATCTGCCTGCTACCTGAGAGAGCCTCTGCCTACAACAACCGGGCTCAAGCGCCGGAGGCTCCA
GGGGGATGTAGCAGGCGCCCTGGAGGACTTGGAGCGCGCAGTGACGCTGAGCGGCGGCCAGGGTCGCCGCCG
CCCGCCAGAGCTTCGTGCAGCGCGGACTGCTGGCGCGATTGCAAGGCCGAGACGACGACGCCCGCAGGGAC
TTCGAGCAGGCAGCGCGACTGGGCAGCCCCTTCGCGCGCGCGCCAGCTGGTGCTGCTCAACCCGTACGCCGCGC
GCTGTGCAACCGCATGCTGGCCGACATGATGGGGCAGCTACGCGCGCCCCAGTAACGGGCGCTGA
```

```
ATGGGGACCCCTAATGACCAGGCTGTGCTGCAGGCCATCTTCAATCCAAATACCCCCTTTGGTGATGTGAT
TGACCTGGACCTGGAAGAGGCCAAGAAAGAGGATGAGGATGGGGTGTTCCCCCAGGAACAATTAGAACAGT
CCAAGGCCCTGGAACTGCAGGGGGTGAGAGCTGCTGAAGCTGGGGATCTGCACACAGCCCTGGAAAAGTTT
GGCCAGGCTATCTGCCTGCTGCCTGAGAGAGCCTCTGCCTACAACAACAGAGCCCAGGCTAGAAGGCTGCA
GGGGGATGTGGCTGGAGCCCTGGAGGATCTGGAAAGAGCTGTGACCCTGTCTGGGGGCCAGGGCAGAGCTG
CCAGACAGTCCTTTGTGCAGAGAGGCCTGCTGGCCAGACTGCAGGGCAGGGATGATGATGCCAGAAGAGAC
TTTGAGCAGGCTGCCAGGCTGGGCAGCCCCTTTGCTAGAAGGCAACTGGTGCTGCTGAACCCCTATGCTGC
CCTGTGCAACAGAATGCTGGCTGACATGATGGGGCAGCTGAGAGCCCCCTCCAATGGCAGATGA
```

Figure 5.22:   The proposed codon optimised coding sequence of *cgr*Ttc36 without CpG islands. A: The *cgr*Ttc36-coding sequence before manual optimisation. Following features are highlighted: CpG Islands (shaded in dark red, white letters); Branch sites (medium grey) and branch points (underlined); 3'-slice sites (dark grey); Cirbp binding sites (light red background, medium red letters); Rbm3 binding sites (medium red background, light red letters). B: Coding sequence after manual optimisation.

Furthermore, other interactors of *cgr*Ttc36 should be revealed to enhance the background of knowledge of the scarcely documented tetratricopeptide repeat-containing protein and ensuring a consistent growth promoting effect by co-expression. Transcripts such as *cgr*Snord78, which were proven to increase the cell specific productivity, should be applied in combination with *cgr*Ttc36 to achieve the maximal recombinant protein yield. For this aim, the intronic sequence of *cgr*Snord78 needs further optimisation as well.

Due to the reduced splicing efficiency of wild-type *cgr*Snord78 intron compared to intron A within CV121 (Figure 4.35), the consensus splice sites need to be optimised. In contrast, the spacing between *cgr*Snord78 and the 3' splice site was optimally evolved towards approx. 70 nucleotides (Hirose and Steitz 2001; Hirose *et al.* 2003). Here, there is no use for optimisation, whereat changes in length of the intronic sequence upstream of the 3' splice site would certainly impair splicing. Thus, the optimisation of the consensus sequences of splice sites, branch points as well as polyT could rather increase splicing efficiency. Hence, stronger splice signals for the intronic *cgr*Snord78 could certainly improve the availability of this snoRNA and would minimise alternative splice moments. Additionally, the predicted binding site of *cgr*-miR-21-5p could impair the proper splicing of Intron$_{cgrSnord78}$ and thus the disruption would be necessary as well.

The application of *cgr*Snord78 in its intronic presence enables the combination with coding sequences towards protein expression directed by a single promoter, which represents a tremendous advantage compared to coding RNA species. Nevertheless, this combination needs to be elucidated, since it could not be revealed whether the co-expression of *cgr*Ttc36 and *cgr*Snord78 represented an incompatible as well as inhibitory combination or the *cgr*Snord78 intron structurally impaired the expression of *cgr*Ttc36 (Figure 4.49 B). The latter effect could be mediated by inhibitory hairpin structures putatively occur in presence of the *cgr*Snord78 intron with *cgr*Ttc36 together. Nevertheless, such hairpins could not be revealed by Mfold web tool (Zuker 2003). Thus, the functional interplay between *cgr*Ttc36 and *cgr*Snord78 or the assumed non-sense splicing of *cgr*Ttc36 (Figure 5.19) are more likely. Combining the putative alternative splicing ability of the employed IRES$_{FMDV}$ sequence (Chapter 5.3.2.4) and the possible binding of *cgr*U2.1 snRNA by *cgr*Snord78 (Figure 5.18), the latter could influence the branch point recognition resulting in *cgr*Ttc36 coding sequence exclusion by splicing. Avoiding or even improving the IRES$_{FMDV}$ sequence towards abolished splice signals could enable the co-expression of *cgr*Ttc36 and *cgr*Snord78. Thus, the coding sequence of *cgr*Ttc36 needs to be optimised in the same direction as mentioned above. Nevertheless, a detailed view on the co-expression of *cgr*Ttc36 and *cgr*Snord78 could indicate further insights whether a functional or intronic interference led to the observed effect (Figure 4.49 B).

Utterly and next to quantity, the quality (glycosylations, post-translational modifications in general) of the produced biopharmaceutical protein is of major importance. Therefore, the impact of the overexpression of *cgr*Ttc36 as well as *cgr*Snord78 should be further revealed, which could be not performed within this thesis. Regarding the overexpression of *cgr*Ttc36, the dolichol-phosphate mannose (DPM; mannosylphosphodolichol) synthase could be modulated at least. The unbound catalytic subunit DPM1 was reported to be rapidly ubiquitinated by *cgr*Chip in CHO-K1 (Ashida *et al.* 2006). Thus, the overexpression of *cgr*Ttc36 could stabilise the DPM1 and therefore the mannosyl donor formation. Alternatively, the direct TPR-mediated binding of *cgr*Ttc36 to DPM1 could disrupt the complex formation with DPM2 or DPM3 and thus the proper endoplasmic translocation (Maeda *et al.* 1998). Hence, a putative DPM1-inhibition could induce N-gylcosylation-deficiency similarily to DPM2-deficient CHO B4-2-1 (Lec15) cells (Jones *et al.* 2010), which could be restored by DPM3-overexpression (Maeda *et al.* 2000). Interestingly, *hsa*SNORD115 was found to be involved in alternative splicing of DPM2 towards non-sense translation (Kishore *et al.* 2010), which could be inhibited by the introduction of CV146 or related CV001/CV121-based vectors (Figure 5.11). Overall, the overexpression of *cgr*Ttc36 putatively possesses an impact on the mannosyl donor formation and the overall N-glycosylation, whether in a stimulating or inhibitory manner.

On the other hand, *cgr*Snord78 was predicted to have no targets within the N-glycosylation pathway (BLAST; Altschul *et al.* 1990). Using *mmu*- or *hsa*-miR-10a and DIANA-microT v3.0 (Maragkakis *et al.* 2009), no relevant targets could be found either. Therefore, there should be further effort for experimental elucidation and the influence of *cgr*Snord78 on N-glycosylation, since this C/D box snoRNA possessed the highest potential in increasing the volumetric productivity in K20-3 producer cell line (Figure 4.31).

Taken together, the separate overexpression of native *cgr*Ttc36 or *cgr*Snord78 revealed beneficial effects for biopharmaceutical protein production. On the one hand, overexpressing *cgr*Ttc36 was advantageous regarding integral viable cell density as well as abolishment of cell aggregation. The latter effect is surely important for reducing costs of detergent additives, downstream processing as well as post-production cleaning. On the other hand, *cgr*Snord78 efficiently increased the cell specific productivity. Therefore, establishing the combined non-impairing overexpression of *cgr*Ttc36 and *cgr*Snord78 could generate the wished "egg-laying, milk-bearing woolly sow" in biopharmaceutical protein production.

6 Conclusion and perspectives

Within the present memorandum, a high cell density-growing CHO-S clone was established and possibly participating cellular factors were identified. Various very promising factors were validated in CHO-S cell lines finally resulting in two functionally differing biomolecules improving the overall growth characteristics (*cgr*Ttc36) or basal cell specific hIgG productivity (*cgr*Snord78), respectively. This summary reflects the conducted common theme throughout this study towards economical efficiency of a production cell line ultimately heading in reducing the ecological footprint in biomanufacturing as well. In total, the achieved results can sustainably reduce the expenses in production and downstream processing.

Fundamental methods for selecting genetically altered cells towards high cell density-growth and reduced cell aggregation were established. Naturally, the applied procedure could not stand for high scientific art, but was highly efficient in output, work capacity as well as economic effort. Unless other research groups that worked on the same problem in parallel, this plain method produced one superior and three highly cell density-decoupled clonal cell lines. Nevertheless, the transfection of CHO-S-E400/1.104 and non-presented parallel mutation as well as selection of other CHO-based production cell lines revealed a less beneficial outcome in productivity (Figure 4.13; data not shown). Even this characteristic was not desired in the thesis' task, the findings show the incompatibility of the established method with optimising production cell lines in general. Due to the genomic as well as phenotypic stability of the selected clonal cell line and possible changes in product quality by chemical mutation, the subsequent characterisation would be too labour-intensive. Hence, deregulated factors were aimed to be applied in unaltered cell lines once validated.

The subsequent external transcriptomic analysis by HiSeq2000 next-generation sequencing revealed a rather inconsistent pattern compared to independent internal qRT-PCR studies (factor verification, Figure 4.18). The relevant data and background for the further validation by transfection was chosen to be the internal achieved qRT-PCR results. Regarding the samples for the transcriptomic analysis, the shipment was performed properly, possibly in contrast to the followed storage conditions. Hence, if the pellets partly thawed, the results would not have been reliable anymore due to rapid mRNA-degrading processes. Therefore, the thread became inconsistent. Nonetheless, the decision focussing on the qRT-PCR results was right and resulted in six possible targets (Figure 4.19).

Again, the separate expression of the upregulated factors revealed the non-triviality in cellular research. Especially, the *C. griseus* homologue of tumour-promoting protein TPRp (*cgr*TPRp) was surprisingly ineffective to increase the viable cell density or growth rate (Chapter 4.3.1). Unfortunately, the effect in producer cell lines was not determined, which could possibly show a beneficial or also inhibitory effect on recombinant protein production. This remains to be revealed. Nevertheless, the latter was established for *cgr*C10orf93-like/*cgr*Ttc40-like (LOC100759461), this formerly undescribed protein resulted in grow arrest or irrelevant change in hIgG titer, even the cell specific productivity was slightly increased (Chapter 4.3.2). Regardless the negative aspects of *cgr*C10orf93-like/*cgr*Ttc40-like, the cell aggregation could be decreased massively in K20-3 producer cell line and thus possess a common feature with the structurally different protein *cgr*Ttc36.

The long non-coding RNA species *cgr*Gas5 was assumed to be evolutionary truncated absent of 5'- as well as 3'-ends observed in *M. musculus* and *R. norvegicus*. This rather courageous presumption was rectified by recently sequenced genome data from various CHO cell lines (Chapter 5.4.1). Hence, the structure and C/D box snoRNA-content of effective full length *cgr*Gas5 is comparable with the other two rodent Gas5. Therefore, the hamster sequences up- as well as downstream of cgrGas5$_{2029bp}$ were not characterised in addition to the intronic C/D box snoRNAs beyond *cgr*Snord78 and options for further findings are obvious. Nonetheless, the cytostatic portion in Gas5 could be assumed within these absent flanking regions. The effect of the already tested *cgr*Gas5 variants on the productivity could be topic for further research as well. At the end, the *cgr*Gas5-derived intronic *cgr*Snord78 could be found as a powerful component to increase the cell specific productivity by various suggested mechanisms (Chapter 5.4.2).

Compared to *cgr*Snord78, the overexpression of *cgr*Ttc36 resulted in an increased integral viable cell density and inhibited cell aggregation surprisingly leading to the halved size of the cell pellet per identical cell density compared to the control. In contrast, the overexpression of the native intronically placed *cgr*Snord78 resulted in a massive increase in cell specific productivity, without affecting the cell density. Hence, each factor could be applied on the respective demand on the beneficial characteristic and thus improving the protein production capacity of a cell. Highly aggregating and producing cell lines could therefore be supertransfected with the coding sequence of the native or an optimised *cgr*Ttc36 to improve the yield as well as efficiency of downstream purification processes. On the other hand, the overexpression of *cgr*Snord78 could improve cell lines with poor cell specific productivities. Obviously, both factors are not generally applicable and each cell line would possess another cellular component content as well as pathways equilibrium. Hence, the applicability needs to be tested for each cell line. In addition, the already positively tested factors *cgr*Snord78 and *cgr*Ttc36 could be further optimised regarding their stability, specificity and avoiding inhibitory effects (Chapters 5.5). Nevertheless, the present native biomolecules showed satisfying characteristics without need for factor, but certain steps in vector optimisation.

In this thesis, the foundations for future research on recent functionally unrevealed factors were laid. Hence, the present work represents the first step towards the application of the genuine biomolecules *cgr*Ttc36 and *cgr*Snord78 in biopharmaceutical protein production. Further factor optimisation could be necessary for certain applications. Nonetheless, the ultimate aim of this thesis was fulfilled with regard to obtain newly discovered cellular factors achieving an increase in recombinant protein production.

7 References

Abrams, J. M.; White, K.; Fessler, L. I.; Steller, H. (1993): Programmed cell death during Drosophila embryogenesis. *Development* 117 (1), S. 29–43.

Aitken, Alastair; Learmonth, Michele (2002): Protein identification by in-gel digestion and mass spectrometric analysis. *Mol. Biotechnol.* 20 (1), S. 95–97.

Althoff, Kristina; Beckers, Anneleen; Odersky, Andrea; Mestdagh, Pieter; Köster, Johannes; Bray, Isabella M. *et al.* (2013): MiR-137 functions as a tumor suppressor in neuroblastoma by downregulating KDM1A. *Int. J. Cancer* 133 (5), S. 1064–1073.

Altschul, S. F.; Gish, W.; Miller, W.; Myers, E. W.; Lipman, D. J. (1990): Basic local alignment search tool. *J. Mol. Biol.* 215 (3), S. 403–410.

Amin, J.; Ananthan, J.; Voellmy, R. (1988): Key features of heat shock regulatory elements. *Mol. Cell. Biol.* 8 (9), S. 3761–3769.

Anckar, Julius; Hietakangas, Ville; Denessiouk, Konstantin; Thiele, Dennis J.; Johnson, Mark S.; Sistonen, Lea (2006): Inhibition of DNA binding by differential sumoylation of heat shock factors. *Mol. Cell. Biol.* 26 (3), S. 955–964.

Andreou, Artemisia M.; Tavernarakis, Nektarios (2009): SUMOylation and cell signalling. *Biotechnol J* 4 (12), S. 1740–1752.

Andrew, Charles D.; Warwicker, Jim; Jones, Gareth R.; Doig, Andrew J. (2002): Effect of Phosphorylation on α-Helix Stability as a Function of Position †. *Biochemistry* 41 (6), S. 1897–1905.

Ansari, Khairul I.; Hussain, Imran; Das, Hriday K.; Mandal, Subhrangsu S. (2009): Overexpression of human histone methylase MLL1 upon exposure to a food contaminant mycotoxin, deoxynivalenol. *FEBS J.* 276 (12), S. 3299–3307.

Appaiah, Hitesh N.; Goswami, Chirayu P.; Mina, Lida A.; Badve, Sunil; Sledge, George W.; Liu, Yunlong; Nakshatri, Harikrishna (2011): Persistent upregulation of U6:SNORD44 small RNA ratio in the serum of breast cancer patients. *Breast Cancer Res.* 13 (5), S. R86.

Arboleda, Valerie A.; Lee, Hane; Parnaik, Rahul; Fleming, Alice; Banerjee, Abhik; Ferraz-de-Souza, Bruno *et al.* (2012): Mutations in the PCNA-binding domain of CDKN1C cause IMAGe syndrome. *Nat. Genet.* 44 (7), S. 788–792.

Ashida, Hisashi; Maeda, Yusuke; Kinoshita, Taroh (2006): DPM1, the catalytic subunit of dolichol-phosphate mannose synthase, is tethered to and stabilized on the endoplasmic reticulum membrane by DPM3. *J. Biol. Chem.* 281 (2), S. 896–904.

Ast, Gil (2004): How did alternative splicing evolve? In: *Nat Rev Genet* 5 (10), S. 773–782.

Bahr, Scott M.; Borgschulte, Trissa; Kayser, Kevin J.; Lin, Nan (2009): Using microarray technology to select housekeeping genes in Chinese hamster ovary cells. *Biotechnol. Bioeng.* 104 (5), S. 1041–1046.

Baik, Jong Youn; Lee, Gyun Min (2010): A DIGE approach for the assessment of differential expression of the CHO proteome under sodium butyrate addition: Effect of Bcl-x(L) overexpression. *Biotechnol. Bioeng.* 105 (2), S. 358–367.

Ballinger, C. A.; Connell, P.; Wu, Y.; Hu, Z.; Thompson, L. J.; Yin, L. Y.; Patterson, C. (1999): Identification of CHIP, a novel tetratricopeptide repeat-containing protein that interacts with heat shock proteins and negatively regulates chaperone functions. *Mol. Cell. Biol.* 19 (6), S. 4535–4545.

Barron, N.; Kumar, N.; Sanchez, N.; Doolan, P.; Clarke, C.; Meleady, P. *et al.* (2011): Engineering CHO cell growth and recombinant protein productivity by overexpression of miR-7. *J. Biotechnol.* 151 (2), S. 204–211.

Baycin-Hizal, Deniz; Tabb, David L.; Chaerkady, Raghothama; Chen, Lily; Lewis, Nathan E.; Nagarajan, Harish *et al.* (2012): Proteomic analysis of Chinese hamster ovary cells. *J. Proteome Res.* 11 (11), S. 5265–5276.

Beckmann, T. F.; Krämer, O.; Klausing, S.; Heinrich, C.; Thüte, T.; Büntemeyer, H. *et al.* (2012): Effects of high passage cultivation on CHO cells: a global analysis. *Appl. Microbiol. Biotechnol.* 94 (3), S. 659–671.

Bentley, Cornelia A.; Bazirgan, Omar A.; Graziano, James; Holmes, Evan M.; Smider, Vaughn V. (2013): Arrayed antibody library technology for therapeutic biologic discovery. *Methods*.

Bettermann, Kira; Benesch, Martin; Weis, Serge; Haybaeck, Johannes (2012): SUMOylation in carcinogenesis. *Cancer Lett.* 316 (2), S. 113–125.

Bhan, Arunoday; Hussain, Imran; Ansari, Khairul I.; Kasiri, Sahba; Bashyal, Aarti; Mandal, Subhrangsu S. (2013): Antisense transcript long noncoding RNA (lncRNA) HOTAIR is transcriptionally induced by estradiol. *J. Mol. Biol.* 425 (19), S. 3707–3722.

Binda, Olivier; LeRoy, Gary; Bua, Dennis J.; Garcia, Benjamin A.; Gozani, Or; Richard, Stéphane (2010): Trimethylation of histone H3 lysine 4 impairs methylation of histone H3 lysine 9: regulation of lysine methyltransferases by physical interaction with their substrates. *Epigenetics* 5 (8), S. 767–775.

Birukova, Anna A.; Tian, Xinyong; Tian, Yufeng; Higginbotham, Katherine; Birukov, Konstantin G. (2013): Rap-afadin axis in control of Rho signaling and endothelial barrier recovery. *Mol. Biol. Cell* 24 (17), S. 2678–2688.

Birzele, Fabian; Schaub, Jochen; Rust, Werner; Clemens, Christoph; Baum, Patrick; Kaufmann, Hitto *et al.* (2010): Into the unknown: expression profiling without genome sequence information in CHO by next generation sequencing. *Nucleic Acids Res.* 38 (12), S. 3999–4010.

Bi, Yueyang; Han, Yong; Bi, Haiyang; Gao, Fuquan; Wang, Xiaozhi (2013): miR-137 impairs the proliferative and migratory capacity of human non-small cell lung cancer cells by targeting paxillin. *Hum. Cell*.

Bliss, Katherine T.; Chu, Miensheng; Jones-Weinert, Colin M.; Gregorio, Carol C. (2013): Investigating lasp-2 in cell adhesion: new binding partners and roles in motility. *Mol. Biol. Cell* 24 (7), S. 995–1006.

Blom, N.; Gammeltoft, S.; Brunak, S. (1999): Sequence and structure-based prediction of eukaryotic protein phosphorylation sites. *J. Mol. Biol.* 294 (5), S. 1351–1362.

Blom, Nikolaj; Sicheritz-Pontén, Thomas; Gupta, Ramneek; Gammeltoft, Steen; Brunak, Søren (2004): Prediction of post-translational glycosylation and phosphorylation of proteins from the amino acid sequence. *Proteomics* 4 (6), S. 1633–1649.

Böhm, Ernst; Voglauer, Regina; Steinfellner, Willi; Kunert, Renate; Borth, Nicole; Katinger, Hermann (2004): Screening for improved cell performance: selection of subclones with altered production kinetics or improved stability by cell sorting. *Biotechnol. Bioeng.* 88 (6), S. 699–706.

Bort, Juan A. Hernández; Stern, Beate; Borth, Nicole (2010): CHO-K1 host cells adapted to growth in glutamine-free medium by FACS-assisted evolution. *Biotechnol J* 5 (10), S. 1090–1097.

Bortolin, M. L.; Kiss, T. (1998): Human U19 intron-encoded snoRNA is processed from a long primary transcript that possesses little potential for protein coding. *RNA* 4 (4), S. 445–454.

Borys, M. C.; Linzer, D. I.; Papoutsakis, E. T. (1994): Ammonia affects the glycosylation patterns of recombinant mouse placental lactogen-I by chinese hamster ovary cells in a pH-dependent manner. *Biotechnol. Bioeng.* 43 (6), S. 505–514.

Brameier, Markus; Krings, Andrea; MacCallum, Robert M. (2007): NucPred--predicting nuclear localization of proteins. *Bioinformatics* 23 (9), S. 1159–1160.

Branda, R. F.; Lafayette, A. R.; O'Neill, J. P.; Nicklas, J. A. (1999): The effect of folate deficiency on the hprt mutational spectrum in Chinese hamster ovary cells treated with monofunctional alkylating agents. *Mutat. Res.* 427 (2), S. 79–87.

Brochu, Christian; Cabrita, Miguel A.; Melanson, Brian D.; Hamill, Jeffrey D.; Lau, Rosanna; Pratt, M. A. Christine; McKay, Bruce C. (2013): NF-κB-dependent role for cold-inducible RNA binding protein in regulating interleukin 1β. *PLoS ONE* 8 (2), S. e57426.

Brocker, Chad; Cantore, Miriam; Failli, Paola; Vasiliou, Vasilis (2011): Aldehyde dehydrogenase 7A1 (ALDH7A1) attenuates reactive aldehyde and oxidative stress induced cytotoxicity. *Chem. Biol. Interact.* 191 (1-3), S. 269–277.

Brocker, Chad; Lassen, Natalie; Estey, Tia; Pappa, Aglaia; Cantore, Miriam; Orlova, Valeria V. *et al.* (2010): Aldehyde dehydrogenase 7A1 (ALDH7A1) is a novel enzyme involved in cellular defense against hyperosmotic stress. *J. Biol. Chem.* 285 (24), S. 18452–18463.

Brunet Simioni, M.; Thonel, A. de; Hammann, A.; Joly, A. L.; Bossis, G.; Fourmaux, E. *et al.* (2009): Heat shock protein 27 is involved in SUMO-2/3 modification of heat shock factor 1 and thereby modulates the transcription factor activity. *Oncogene* 28 (37), S. 3332–3344.

Bunai, Keigo; Yamane, Kunio (2005): Effectiveness and limitation of two-dimensional gel electrophoresis in bacterial membrane protein proteomics and perspectives. *J. Chromatogr. B Analyt. Technol. Biomed. Life Sci.* 815 (1-2), S. 227–236.

Burton, Elizabeth R.; Gaffar, Aneesa; Lee, Soo Jin; Adeshuko, Folashade; Whitney, Kathleen D.; Chung, Joon-Yong *et al.* (2011): Downregulation of Filamin A interacting protein 1-like is associated with promoter methylation and induces an invasive phenotype in ovarian cancer. *Mol. Cancer Res.* 9 (8), S. 1126–1138.

Butler, M.; Meneses-Acosta, A. (2012): Recent advances in technology supporting biopharmaceutical production from mammalian cells. *Appl. Microbiol. Biotechnol.* 96 (4), S. 885–894.

Calzado, Marco A.; Sancho, Rocío; Muñoz, Eduardo (2004): Human immunodeficiency virus type 1 Tat increases the expression of cleavage and polyadenylation specificity factor 73-kilodalton subunit modulating cellular and viral expression. *J. Virol.* 78 (13), S. 6846–6854.

Campos, Catarina; Valente, Luísa M.P; Conceição, Luís E.C; Engrola, Sofia; Fernandes, Jorge M.O (2013): Temperature affects methylation of the myogenin putative promoter, its expression and muscle cellularity in Senegalese sole larvae. *epigenetics* 8 (4), S. 389–397.

Canel, M.; Serrels, A.; Frame, M. C.; Brunton, V. G. (2013): E-cadherin-integrin crosstalk in cancer invasion and metastasis. *Journal of Cell Science* 126 (2), S. 393–401.

Cao, Shenglan; Ho, Gay; Lin, Valerie C. L. (2008): Tetratricopeptide repeat domain 9A is an interacting protein for tropomyosin Tm5NM-1. *BMC Cancer* 8 (1), S. 231.

Cao, Shenglan; Iyer, Jayasree K.; Lin, Valerie (2006): Identification of tetratricopeptide repeat domain 9, a hormonally regulated protein. *Biochem. Biophys. Res. Commun.* 345 (1), S. 310–317.

Castellano, Leandro; Stebbing, Justin (2013): Deep sequencing of small RNAs identifies canonical and non-canonical miRNA and endogenous siRNAs in mammalian somatic tissues. *Nucleic Acids Res.* 41 (5), S. 3339–3351.

Cates, Jordan; Graham, Garrett C.; Omattage, Natalie; Pavesich, Elizabeth; Setliff, Ian; Shaw, Jack *et al.* (2011): Sensing the heat stress by Mammalian cells. *BMC Biophys* 4, S. 16.

Chang, Pei-Yun; Hom, Robert A.; Musselman, Catherine A.; Zhu, Li; Kuo, Alex; Gozani, Or *et al.* (2010): Binding of the MLL PHD3 Finger to Histone H3K4me3 Is Required for MLL-Dependent Gene Transcription. *Journal of Molecular Biology* 400 (2), S. 137–144.

Chatterjee, Anuran; Snead, Connie; Yetik-Anacak, Gunay; Antonova, Galina; Zeng, Jingmin; Catravas, John D. (2008): Heat shock protein 90 inhibitors attenuate LPS-induced endothelial hyperpermeability. *Am. J. Physiol. Lung Cell Mol. Physiol.* 294 (4), S. L755-63.

Chau, B.Nelson; Cheng, Emily H.-Y; Kerr, Douglas A.; Hardwick, J.Marie (2000): Aven, a Novel Inhibitor of Caspase Activation, Binds Bcl-xL and Apaf-1. *Molecular Cell* 6 (1), S. 31–40.

Cheng, Ibis Kc; Tsang, Bruce Ck; Lai, Keng Po; Ching, Arthur Kk; Chan, Anthony Wh; To, Ka-Fai *et al.* (2012): GEF-H1 over-expression in hepatocellular carcinoma promotes cell motility via activation of RhoA signalling. *J. Pathol.*

Chen, L. N.; Wang, Y.; Ma, D. L.; Chen, Y. Y. (2006): Short interfering RNA against the PDCD5 attenuates cell apoptosis and caspase-3 activity induced by Bax overexpression. *Apoptosis* 11 (1), S. 101–111.

Chen, Peifeng; Harcum, Sarah W. (2005): Effects of amino acid additions on ammonium stressed CHO cells. *J. Biotechnol.* 117 (3), S. 277–286.

Chen, S. (1998): Hop as an Adaptor in the Heat Shock Protein 70 (Hsp70) and Hsp90 Chaperone Machinery. *Journal of Biological Chemistry* 273 (52), S. 35194–35200.

Chen, Xiaobing; Chen, Guoyong; Cao, Xinguang; Zhou, Yudong; Yang, Tiejun; Wei, Sidong (2013a): Downregulation of BTG3 in non-small cell lung cancer. *Biochem. Biophys. Res. Commun.* 437 (1), S. 173–178.

Chen, Yuan-Ling; Liang, Hui-Lin; Ma, Xing-Liang; Lou, Su-Lin; Xie, Yong-Yao; Liu, Zhen-Lan *et al.* (2013b): An efficient rice mutagenesis system based on suspension-cultured cells. *J Integr Plant Biol* 55 (2), S. 122–130.

Chong, William P. K.; Reddy, Satty G.; Yusufi, Faraaz N. K.; Lee, Dong-Yup; Wong, Niki S. C.; Heng, Chew Kiat *et al.* (2010): Metabolomics-driven approach for the improvement of Chinese hamster ovary cell growth: overexpression of malate dehydrogenase II. *J. Biotechnol.* 147 (2), S. 116–121.

Chuang, Jian-Ying; Chang, Wen-Chang; Hung, Jan-Jong (2011): Hydrogen peroxide induces Sp1 methylation and thereby suppresses cyclin B1 via recruitment of Suv39H1 and HDAC1 in cancer cells. *Free Radic. Biol. Med.* 51 (12), S. 2309–2318.

Ciobanasu, Corina; Faivre, Bruno; Le Clainche, Christophe (2013): Integrating actin dynamics, mechanotransduction and integrin activation: the multiple functions of actin binding proteins in focal adhesions. *Eur. J. Cell Biol.* 92 (10-11), S. 339–348.

Clark, David; Mao, Li (2012): Cancer biomarker discovery: lectin-based strategies targeting glycoproteins. *Dis. Markers* 33 (1), S. 1–10.

Clarke, Colin; Henry, Michael; Doolan, Padraig; Kelly, Shane; Aherne, Sinead; Sanchez, Noelia *et al.* (2012): Integrated miRNA, mRNA and protein expression analysis reveals the role of post-transcriptional regulation in controlling CHO cell growth rate. *BMC Genomics* 13, S. 656.

Collins, Samuel L.; Hervé, Rodolphe; Keevil, C. W.; Blaydes, Jeremy P.; Webb, Jeremy S. (2011): Down-regulation of DNA mismatch repair enhances initiation and growth of neuroblastoma and brain tumour multicellular spheroids. *PLoS ONE* 6 (12), S. e28123.

Colzani, Mara; Schütz, Frédéric; Potts, Alexandra; Waridel, Patrice; Quadroni, Manfredo (2008): Relative protein quantification by isobaric SILAC with immonium ion splitting (ISIS). *Mol. Cell Proteomics* 7 (5), S. 927–937.

Cooper, J. A.; King, C. S. (1986): Dephosphorylation or antibody binding to the carboxy terminus stimulates pp60c-src. *Mol. Cell. Biol.* 6 (12), S. 4467–4477.

Cost, Gregory J.; Freyvert, Yevgeniy; Vafiadis, Annamaria; Santiago, Yolanda; Miller, Jeffrey C.; Rebar, Edward *et al.* (2010): BAK and BAX deletion using zinc-finger nucleases yields apoptosis-resistant CHO cells. *Biotechnol. Bioeng.* 105 (2), S. 330–340.

Cottrell, J. S. (1994): Protein identification by peptide mass fingerprinting. *Pept. Res.* 7 (3), S. 115–124.

Courtes, Franck C.; Gu, Chen; Wong, Niki S.C; Dedon, Peter C.; Yap, Miranda G.S; Lee, Dong-Yup (2014): 28S rRNA is inducibly pseudouridylated by the mTOR pathway translational control in CHO cell cultures. *Journal of Biotechnology* 174, S. 16–21.

Courtneidge, S. A. (1985): Activation of the pp60c-src kinase by middle T antigen binding or by dephosphorylation. *EMBO J.* 4 (6), S. 1471–1477.

Cozzi, R.; Ricordy, R.; Aglitti, T.; Gatta, V.; Perticone, P.; Salvia, R. de (1997): Ascorbic acid and beta-carotene as modulators of oxidative damage. *Carcinogenesis* 18 (1), S. 223–228.

Crawford, Brendan D.; Hess, Jay L. (2006): MLL core components give the green light to histone methylation. *ACS Chem. Biol.* 1 (8), S. 495–498.

Cress, A. E.; Majda, J. A.; Glass, J. R.; Stringer, D. E.; Gerner, E. W. (1990): Alteration of cellular adhesion by heat shock. *Exp. Cell Res.* 190 (1), S. 40–46.

D'Ambrogio, Andrea; Nagaoka, Kentaro; Richter, Joel D. (2013): Translational control of cell growth and malignancy by the CPEBs. *Nat. Rev. Cancer* 13 (4), S. 283–290.

Dames, S. A. (2010): Structural Basis for the Association of the Redox-sensitive Target of Rapamycin FATC Domain with Membrane-mimetic Micelles. *Journal of Biological Chemistry* 285 (10), S. 7766–7775.

Davis, R.; Schooley, K.; Rasmussen, B.; Thomas, J.; Reddy, P. (2000): Effect of PDI overexpression on recombinant protein secretion in CHO cells. *Biotechnol. Prog.* 16 (5), S. 736–743.

Dean, Jason; Reddy, Pranhitha (2013): Metabolic analysis of antibody producing CHO cells in fed-batch production. *Biotechnol. Bioeng.*

Demonacos, C.; Krstic-Demonacos, M.; La Thangue, N. B. (2001): A TPR motif cofactor contributes to p300 activity in the p53 response. *Mol. Cell* 8 (1), S. 71–84.

Deng, Boya; Zhao, Yang; Gou, Wenfeng; Chen, Shuo; Mao, Xiaoyun; Takano, Yasuo; Zheng, Huachuan (2013): Decreased expression of BTG3 was linked to carcinogenesis, aggressiveness, and prognosis of ovarian carcinoma. *Tumour Biol.* 34 (5), S. 2617–2624.

Deng, Xiaodi A.; Norris, Andrea; Panaviene, Zivile; Moncman, Carole L. (2008): Ectopic expression of LIM-nebulette (LASP2) reveals roles in cell migration and spreading. *Cell Motil. Cytoskeleton* 65 (10), S. 827–840.

Dere, Edward; Forgacs, Agnes L.; Zacharewski, Timothy R.; Burgoon, Lyle D. (2011): Genome-wide computational analysis of dioxin response element location and distribution in the human, mouse, and rat genomes. *Chem. Res. Toxicol.* 24 (4), S. 494–504.

Derouazi, M.; Martinet, D.; Besuchet Schmutz, N.; Flaction, R.; Wicht, M.; Bertschinger, M. *et al.* (2006): Genetic characterization of CHO production host DG44 and derivative recombinant cell lines. *Biochem. Biophys. Res. Commun.* 340 (4), S. 1069–1077.

Deryusheva, Svetlana; Choleza, Maria; Barbarossa, Adrien; Gall, Joseph G.; Bordonné, Rémy (2012): Post-transcriptional modification of spliceosomal RNAs is normal in SMN-deficient cells. *RNA* 18 (1), S. 31–36.

Dhyani, Anamika; Duarte, Adriana S. S.; Machado-Neto, João A.; Favaro, Patricia; Ortega, Manoela Marques; Olalla Saad, Sara T. (2012): ANKHD1 regulates cell cycle progression and proliferation in multiple myeloma cells. *FEBS Lett.* 586 (24), S. 4311–4318.

Di Cello, Francescopaolo; Shin, James; Harbom, Kirsten; Brayton, Cory (2013): Knockdown of HMGA1 inhibits human breast cancer cell growth and metastasis in immunodeficient mice. *Biochem. Biophys. Res. Commun.* 434 (1), S. 70–74.

Doe, Christine M.; Relkovic, Dinko; Garfield, Alastair S.; Dalley, Jeffrey W.; Theobald, David E. H.; Humby, Trevor et al. (2009): Loss of the imprinted snoRNA mbii-52 leads to increased 5htr2c pre-RNA editing and altered 5HT2CR-mediated behaviour. Hum. Mol. Genet. 18 (12), S. 2140–2148.

Dönmez, Gizem; Hartmuth, Klaus; Lührmann, Reinhard (2004): Modified nucleotides at the 5' end of human U2 snRNA are required for spliceosomal E-complex formation. RNA 10 (12), S. 1925–1933.

Doolan, Padraig; Barron, Niall; Kinsella, Paula; Clarke, Colin; Meleady, Paula; O'Sullivan, Finbarr et al. (2012): Microarray expression profiling identifies genes regulating sustained cell specific productivity (S-Qp) in CHO K1 production cell lines. Biotechnol J 7 (4), S. 516–526.

Doolan, Padraig; Meleady, Paula; Barron, Niall; Henry, Michael; Gallagher, Ross; Gammell, Patrick et al. (2010): Microarray and proteomics expression profiling identifies several candidates, including the valosin-containing protein (VCP), involved in regulating high cellular growth rate in production CHO cell lines. Biotechnol. Bioeng. 106 (1), S. 42–56.

Dorai, Haimanti; Ellis, Dawn; Keung, Yun Seung; Campbell, Marguerite; Zhuang, Minhong; Lin, Chengbin; Betenbaugh, Michael J. (2010): Combining high-throughput screening of caspase activity with anti-apoptosis genes for development of robust CHO production cell lines. Biotechnol. Prog. 26 (5), S. 1367–1381.

Dorai, Haimanti; Kyung, Yun Seung; Ellis, Dawn; Kinney, Cherylann; Lin, Chengbin; Jan, David et al. (2009): Expression of anti-apoptosis genes alters lactate metabolism of Chinese Hamster Ovary cells in culture. Biotechnol. Bioeng. 103 (3), S. 592–608.

Dou, Yali; Milne, Thomas A.; Ruthenburg, Alexander J.; Lee, Seunghee; Lee, Jae Woon; Verdine, Gregory L. et al. (2006): Regulation of MLL1 H3K4 methyltransferase activity by its core components. Nat. Struct. Mol. Biol. 13 (8), S. 713–719.

Dreesen, Imke A. J.; Fussenegger, Martin (2011): Ectopic expression of human mTOR increases viability, robustness, cell size, proliferation, and antibody production of chinese hamster ovary cells. Biotechnol. Bioeng. 108 (4), S. 853–866.

Dresios, John; Aschrafi, Armaz; Owens, Geoffrey C.; Vanderklish, Peter W.; Edelman, Gerald M.; Mauro, Vincent P. (2005): Cold stress-induced protein Rbm3 binds 60S ribosomal subunits, alters microRNA levels, and enhances global protein synthesis. Proc. Natl. Acad. Sci. U.S.A. 102 (6), S. 1865–1870.

Duong-Thi, Minh-Dao; Bergström, Maria; Fex, Tomas; Svensson, Susanne; Ohlson, Sten; Isaksson, Roland (2013): Weak Affinity Chromatography for Evaluation of Stereoisomers in Early Drug Discovery. J Biomol Screen.

Dutca, Laura M.; Gallagher, Jennifer E. G.; Baserga, Susan J. (2011): The initial U3 snoRNA:pre-rRNA base pairing interaction required for pre-18S rRNA folding revealed by in vivo chemical probing. Nucleic Acids Res. 39 (12), S. 5164–5180.

Dutton, R. L.; Scharer, J. M.; Moo-Young, M. (1998): Descriptive parameter evaluation in mammalian cell culture. Cytotechnology 26 (2), S. 139–152.

Endo, Mitsuharu; Yamashita, Toshihide (2009): Inactivation of Ras by p120GAP via focal adhesion kinase dephosphorylation mediates RGMa-induced growth cone collapse. J. Neurosci. 29 (20), S. 6649–6662.

Ernoult-Lange, Michèle; Wilczynska, Ania; Harper, Maryannick; Aigueperse, Christelle; Dautry, François; Kress, Michel; Weil, Dominique (2009): Nucleocytoplasmic traffic of CPEB1 and accumulation in Crm1 nucleolar bodies. Mol. Biol. Cell 20 (1), S. 176–187.

Ernst, Wolfgang; Trummer, Evelyn; Mead, Jennifer; Bessant, Conrad; Strelec, Harald; Katinger, Hermann; Hesse, Friedemann (2006): Evaluation of a genomics platform for cross-species transcriptome analysis of recombinant CHO cells. Biotechnol J 1 (6), S. 639–650.

Evdokimovskaya, Yulia; Skarga, Yuri; Vrublevskaya, Veronika; Morenkov, Oleg (2010): Secretion of the heat shock proteins HSP70 and HSC70 by baby hamster kidney (BHK-21) cells. *Cell Biol. Int.* 34 (10), S. 985–990.

Fan, Dorothy Ngo-Yin; Tsang, Felice Ho-Ching; Tam, Aegean Hoi-Kam; Au, Sandy Leung-Kuen; Wong, Carmen Chak-Lui; Wei, Lai *et al.* (2013): Histone lysine methyltransferase, suppressor of variegation 3-9 homolog 1, promotes hepatocellular carcinoma progression and is negatively regulated by microRNA-125b. *Hepatology* 57 (2), S. 637–647.

Ferrando, I. M.; Chaerkady, R.; Zhong, J.; Molina, H.; Jacob, H. K. C.; Herbst-Robinson, K. *et al.* (2012): Identification of Targets of c-Src Tyrosine Kinase by Chemical Complementation and Phosphoproteomics. *Molecular & Cellular Proteomics* 11 (8), S. 355–369.

Figueroa, Bruno; Chen, Sulin; Oyler, George A.; Hardwick, J. Marie; Betenbaugh, Michael J. (2004): Aven and Bcl-xL enhance protection against apoptosis for mammalian cells exposed to various culture conditions. *Biotechnol. Bioeng.* 85 (6), S. 589–600.

Filipczak, Piotr Teodor; Piglowski, Wojciech; Glowala-Kosinska, Magdalena; Krawczyk, Zdzislaw; Scieglinska, Dorota (2012): HSPA2 overexpression protects V79 fibroblasts against bortezomib-induced apoptosis. *Biochem. Cell Biol.* 90 (2), S. 224–231.

Fiore, Mario; Zanier, Romina; Degrassi, Francesca (2002): Reversible G(1) arrest by dimethyl sulfoxide as a new method to synchronize Chinese hamster cells. *Mutagenesis* 17 (5), S. 419–424.

Flintoff, W. F.; Davidson, S. V.; Siminovitch, L. (1976): Isolation and partial characterization of three methotrexate-resistant phenotypes from Chinese hamster ovary cells. *Somatic Cell Genet.* 2 (3), S. 245–261.

Fornace, Albert J.; Alamo, Isaac; Hollander, M.Christine; Lamoreaux, Etienne (1989): Induction of heat shock protein transcripts and B2 transcripts by various stresses in Chinese hamster cells. *Experimental Cell Research* 182 (1), S. 61–74.

Fortes, P.; Bilbao-Cortés, D.; Fornerod, M.; Rigaut, G.; Raymond, W.; Séraphin, B.; Mattaj, I. W. (1999): Luc7p, a novel yeast U1 snRNP protein with a role in 5' splice site recognition. *Genes Dev.* 13 (18), S. 2425–2438.

Fox, Stephen R.; Tan, Hong Kiat; Tan, Mei Chee; Wong, S. C. Niki C.; Yap, Miranda G. S.; Wang, Daniel I. C. (2005): A detailed understanding of the enhanced hypothermic productivity of interferon-gamma by Chinese-hamster ovary cells. *Biotechnol. Appl. Biochem.* 41 (Pt 3), S. 255–264.

Freitas Junior, Julio Cesar Madureira de; Du Silva, Bárbara Rocher D'Aguiar; Souza, Waldemir Fernandes de; Araújo, Wallace Martins de; Abdelhay, Eliana Saul Furquim Werneck; Morgado-Díaz, José Andrés (2011): Inhibition of N-linked glycosylation by tunicamycin induces E-cadherin-mediated cell-cell adhesion and inhibits cell proliferation in undifferentiated human colon cancer cells. *Cancer Chemother. Pharmacol.* 68 (1), S. 227–238.

Fu, Da-Zhi; Cheng, Ying; He, Hui; Liu, Hai-Yang; Liu, Yong-Feng (2013): PDCD5 expression predicts a favorable outcome in patients with hepatocellular carcinoma. *Int. J. Oncol.* 43 (3), S. 821–830.

Fuscoe, J. C.; O'Neill, J. P.; Peck, R. M.; Hsie, A. W. (1979): Mutagenicity and cytotoxicity of nineteen heterocyclic mustards (ICR compounds) in cultured mammalian cells. *Cancer Res.* 39 (12), S. 4875–4881.

Fu, Szu-Chin; Imai, Kenichiro; Horton, Paul (2011): Prediction of leucine-rich nuclear export signal containing proteins with NESsential. *Nucleic Acids Res.* 39 (16), S. e111.

Fu, Tao; Blei, Andres T.; Takamura, Noriaki; Lin, Tesu; Guo, Danqing; Li, Honglin *et al.* (2004): Hypothermia inhibits Fas-mediated apoptosis of primary mouse hepatocytes in culture. *Cell Transplant* 13 (6), S. 667–676.

Gabanyi, Margaret J.; Adams, Paul D.; Arnold, Konstantin; Bordoli, Lorenza; Carter, Lester G.; Flippen-Andersen, Judith *et al.* (2011): The Structural Biology Knowledgebase: a portal to protein structures, sequences, functions, and methods. *J Struct Funct Genomics* 12 (2), S. 45–54.

Gammell, Patrick; Barron, Niall; Kumar, Niraj; Clynes, Martin (2007): Initial identification of low temperature and culture stage induction of miRNA expression in suspension CHO-K1 cells. *J. Biotechnol.* 130 (3), S. 213–218.

Gaur, Arti B.; Holbeck, Susan L.; Colburn, Nancy H.; Israel, Mark A. (2011): Downregulation of Pdcd4 by mir-21 facilitates glioblastoma proliferation in vivo. *Neuro-oncology* 13 (6), S. 580–590.

Geiger, Tamar; Cox, Juergen; Mann, Matthias (2010): Proteomic changes resulting from gene copy number variations in cancer cells. *PLoS Genet.* 6 (9).

Ge, Junhui; Liu, Huimin; Yu, Yi-Tao (2010): Regulation of pre-mRNA splicing in Xenopus oocytes by targeted 2'-O-methylation. *RNA* 16 (5), S. 1078–1085.

Ghaedi, Kamran; Fujiki, Yukio (2008): Isolation and characterization of novel phenotype CHO cell mutants defective in peroxisome assembly, using ICR191 as a potent mutagenic agent. *Cell Biochem. Funct.* 26 (6), S. 684–691.

Giorgianni, Francesco; Desiderio, Dominic M.; Beranova-Giorgianni, Sarka (2003): Proteome analysis using isoelectric focusing in immobilized pH gradient gels followed by mass spectrometry. *Electrophoresis* 24 (1-2), S. 253–259.

Giovannini, Catia; Gramantieri, Laura; Minguzzi, Manuela; Fornari, Francesca; Chieco, Pasquale; Grazi, Gian Luca; Bolondi, Luigi (2012): CDKN1C/P57 is regulated by the Notch target gene Hes1 and induces senescence in human hepatocellular carcinoma. *Am. J. Pathol.* 181 (2), S. 413–422.

Griffiths-Jones, S.; Saini, H. K.; van Dongen, S.; Enright, A. J. (2007): miRBase: tools for microRNA genomics. *Nucleic Acids Research* 36 (Database), S. D154.

Griffiths-Jones, Sam; Grocock, Russell J.; van Dongen, Stijn; Bateman, Alex; Enright, Anton J. (2006): miRBase: microRNA sequences, targets and gene nomenclature. *Nucleic Acids Res.* 34 (Database issue), S. D140-4.

Grønborg, Mads; Kristiansen, Troels Zakarias; Iwahori, Akiko; Chang, Rubens; Reddy, Raghunath; Sato, Norihiro *et al.* (2006): Biomarker discovery from pancreatic cancer secretome using a differential proteomic approach. *Mol. Cell Proteomics* 5 (1), S. 157–171.

Grudzien-Nogalska, Ewa; Reed, Brent C.; Rhoads, Robert E. (2014): CPEB1 promotes differentiation and suppresses EMT in mammary epithelial cells. *J. Cell. Sci.*

Guduric-Fuchs, Jasenka; O'Connor, Anna; Cullen, Angela; Harwood, Laura; Medina, Reinhold J.; O'Neill, Christina L. *et al.* (2012): Deep sequencing reveals predominant expression of miR-21 amongst the small non-coding RNAs in retinal microvascular endothelial cells. *J. Cell. Biochem.* 113 (6), S. 2098–2111.

Guo (2011): Cell surface heat shock protein 90 modulates prostate cancer cell adhesion and invasion through the integrin-β1/focal adhesion kinase/c-Src signaling pathway. *Oncol Rep* 25 (5).

Gupta, Ramneek; Brunak, Søren (2002): Prediction of glycosylation across the human proteome and the correlation to protein function. *Pac Symp Biocomput*, S. 310–322.

Gupta, Swati; Kim, Se Y.; Artis, Sonja; Molfese, David L.; Schumacher, Armin; Sweatt, J. David *et al.* (2010): Histone methylation regulates memory formation. *J. Neurosci.* 30 (10), S. 3589–3599.

Gurbuxani, Sandeep; Schmitt, Elise; Cande, Celine; Parcellier, Arnaud; Hammann, Arlette; Daugas, Eric *et al.* (2003): Heat shock protein 70 binding inhibits the nuclear import of apoptosis-inducing factor. *Oncogene* 22 (43), S. 6669–6678.

Hackl, Matthias; Jakobi, Tobias; Blom, Jochen; Doppmeier, Daniel; Brinkrolf, Karina; Szczepanowski, Rafael *et al.* (2011): Next-generation sequencing of the Chinese hamster ovary microRNA transcriptome: Identification, annotation and profiling of microRNAs as targets for cellular engineering. *J. Biotechnol.* 153 (1-2), S. 62–75.

Hamajima, Naoki; Johmura, Yoshikazu; Suzuki, Satoshi; Nakanishi, Makoto; Saitoh, Shinji (2013): Increased protein stability of CDKN1C causes a gain-of-function phenotype in patients with IMAGe syndrome. *PLoS ONE* 8 (9), S. e75137.

Hamiel, Christine R.; Pinto, Shanti; Hau, Ann; Wischmeyer, Paul E. (2009): Glutamine enhances heat shock protein 70 expression via increased hexosamine biosynthetic pathway activity. *Am. J. Physiol., Cell Physiol.* 297 (6), S. C1509-19.

Hammond, Stephanie; Kaplarevic, Mihailo; Borth, Nicole; Betenbaugh, Michael J.; Lee, Kelvin H. (2012a): Chinese hamster genome database: an online resource for the CHO community at www.CHOgenome.org. *Biotechnol. Bioeng.* 109 (6), S. 1353–1356.

Hammond, Stephanie; Swanberg, Jeffrey C.; Kaplarevic, Mihailo; Lee, Kelvin H. (2011): Genomic sequencing and analysis of a Chinese hamster ovary cell line using Illumina sequencing technology. *BMC Genomics* 12, S. 67.

Hammond, Stephanie; Swanberg, Jeffrey C.; Polson, Shawn W.; Lee, Kelvin H. (2012b): Profiling conserved microRNA expression in recombinant CHO cell lines using Illumina sequencing. *Biotechnol. Bioeng.* 109 (6), S. 1371–1375.

Han, J.; Lee, J. D.; Bibbs, L.; Ulevitch, R. J. (1994): A MAP kinase targeted by endotoxin and hyperosmolarity in mammalian cells. *Science* 265 (5173), S. 808–811.

Han, Young Kue; Ha, Tae Kwang; Kim, Yeon-Gu; Lee, Gyun Min (2011): Bcl-x(L) overexpression delays the onset of autophagy and apoptosis in hyperosmotic recombinant Chinese hamster ovary cell cultures. *J. Biotechnol.* 156 (1), S. 52–55.

Han, Young Kue; Kim, Yeon-Gu; Kim, Jee Yon; Lee, Gyun Min (2010): Hyperosmotic stress induces autophagy and apoptosis in recombinant Chinese hamster ovary cell culture. *Biotechnol. Bioeng.* 105 (6), S. 1187–1192.

Hecker, Timothy P.; Ding, Qiang; Rege, Tanya A.; Hanks, Steven K.; Gladson, Candece L. (2004): Overexpression of FAK promotes Ras activity through the formation of a FAK/p120RasGAP complex in malignant astrocytoma cells. *Oncogene* 23 (22), S. 3962–3971.

He, Lei; Torres-Lockhart, Kristine; Forster, Nicole; Ramakrishnan, Saranya; Greninger, Patricia; Garnett, Mathew J. *et al.* (2013): Mcl-1 and FBW7 control a dominant survival pathway underlying HDAC and Bcl-2 inhibitor synergy in squamous cell carcinoma. *Cancer Discov* 3 (3), S. 324–337.

Hemmings, B. A. (1997): Akt signaling: linking membrane events to life and death decisions. *Science* 275 (5300), S. 628–630.

Higa-Nakamine, S.; Suzuki, T.; Uechi, T.; Chakraborty, A.; Nakajima, Y.; Nakamura, M. *et al.* (2011): Loss of ribosomal RNA modification causes developmental defects in zebrafish. *Nucleic Acids Research* 40 (1), S. 391–398.

Hinterkörner, Georg; Brugger, Gudrun; Müller, Dethardt; Hesse, Friedemann; Kunert, Renate; Katinger, Hermann; Borth, Nicole (2007): Improvement of the energy metabolism of recombinant CHO cells by cell sorting for reduced mitochondrial membrane potential. *J. Biotechnol.* 129 (4), S. 651–657.

Hirose, T.; Steitz, J. A. (2001): Position within the host intron is critical for efficient processing of box C/D snoRNAs in mammalian cells. *Proc. Natl. Acad. Sci. U.S.A.* 98 (23), S. 12914–12919.

Hirose, Tetsuro; Shu, Mei-Di; Steitz, Joan A. (2003): Splicing-dependent and -independent modes of assembly for intron-encoded box C/D snoRNPs in mammalian cells. *Mol. Cell* 12 (1), S. 113–123.

Hoff, Sylvia; Halbritter, Jan; Epting, Daniel; Frank, Valeska; Nguyen, Thanh-Minh T.; van Reeuwijk, Jeroen *et al.* (2013): ANKS6 is a central component of a nephronophthisis module linking NEK8 to INVS and NPHP3. *Nat. Genet.* 45 (8), S. 951–956.

Hong, Jong Kwang; Kim, Yeon-Gu; Yoon, Sung Kwan; Lee, Gyun Min (2007): Down-regulation of cold-inducible RNA-binding protein does not improve hypothermic growth of Chinese hamster ovary cells producing erythropoietin. *Metab. Eng.* 9 (2), S. 208–216.

Hu, K. Q.; Settleman, J. (1997): Tandem SH2 binding sites mediate the RasGAP-RhoGAP interaction: a conformational mechanism for SH3 domain regulation. *EMBO J.* 16 (3), S. 473–483.

Hunter, S.; Jones, P.; Mitchell, A.; Apweiler, R.; Attwood, T. K.; Bateman, A. *et al.* (2011): InterPro in 2011: new developments in the family and domain prediction database. *Nucleic Acids Research* 40 (D1), S. D306.

Hwang, Sun Ok; Lee, Gyun Min (2009): Effect of Akt overexpression on programmed cell death in antibody-producing Chinese hamster ovary cells. *J. Biotechnol.* 139 (1), S. 89–94.

Irtegun, Sevgi; Wood, Rebecca J.; Ormsby, Angelique R.; Mulhern, Terrence D.; Hatters, Danny M. (2013): Tyrosine 416 is phosphorylated in the closed, repressed conformation of c-Src. *PLoS ONE* 8 (7), S. e71035.

Ivics, Zoltán; Izsvák, Zsuzsanna (2010): The expanding universe of transposon technologies for gene and cell engineering. *Mob DNA* 1 (1), S. 25.

Iwase, Akira; Shen, Ruoqian; Navarro, Daniel; Nanus, David M. (2004): Direct binding of neutral endopeptidase 24.11 to ezrin/radixin/moesin (ERM) proteins competes with the interaction of CD44 with ERM proteins. *J. Biol. Chem.* 279 (12), S. 11898–11905.

Jadhav, Vaibhav; Hackl, Matthias; Bort, Juan A. Hernandez; Wieser, Matthias; Harreither, Eva; Kunert, Renate *et al.* (2012): A screening method to assess biological effects of microRNA overexpression in Chinese hamster ovary cells. *Biotechnol. Bioeng.* 109 (6), S. 1376–1385.

Jakobsson, Magnus E.; Moen, Anders; Bousset, Luc; Egge-Jacobsen, Wolfgang; Kernstock, Stefan; Melki, Ronald; Falnes, Pål Ø. (2013): Identification and characterization of a novel human methyltransferase modulating Hsp70 protein function through lysine methylation. *J. Biol. Chem.* 288 (39), S. 27752–27763.

Jaluria, Pratik; Betenbaugh, Michael; Konstantopoulos, Konstantinos; Shiloach, Joseph (2007): Enhancement of cell proliferation in various mammalian cell lines by gene insertion of a cyclin-dependent kinase homolog. *BMC Biotechnol.* 7, S. 71.

Jang, Kang Won; Lee, Jeong Eun; Kim, Sun Young; Kang, Min-Woong; Na, Myung Hoon; Lee, Choong Sik *et al.* (2011): The C-terminus of Hsp70-interacting protein promotes Met receptor degradation. *J Thorac Oncol* 6 (4), S. 679–687.

Janzer, Andreas; Lim, Soyoung; Fronhoffs, Florian; Niazy, Naima; Buettner, Reinhard; Kirfel, Jutta (2012): Lysine-specific demethylase 1 (LSD1) and histone deacetylase 1 (HDAC1) synergistically repress proinflammatory cytokines and classical complement pathway components. *Biochem. Biophys. Res. Commun.* 421 (4), S. 665–670.

Jaquet, K.; Korte, K.; Schnackerz, K.; Vyska, K.; Heilmeyer, L. M. (1993): Characterization of the cardiac troponin I phosphorylation domain by 31P nuclear magnetic resonance spectroscopy. *Biochemistry* 32 (50), S. 13873–13878.

Jenkins, Nigel; Barron, Niall; Alves, Paula M. (2011): Proceedings of the 21st Annual Meeting of the European Society for Animal Cell Technology (ESACT), Dublin, Ireland, June 7-10, 2009. 1. Aufl. New York: Springer.

Jeon, Min Kyoung; Yu, Da Young; Lee, Gyun Min (2011): Combinatorial engineering of ldh-a and bcl-2 for reducing lactate production and improving cell growth in dihydrofolate reductase-deficient Chinese hamster ovary cells. *Appl. Microbiol. Biotechnol.* 92 (4), S. 779–790.

Jesus, Maria de; Wurm, Florian M. (2011): Manufacturing recombinant proteins in kg-ton quantities using animal cells in bioreactors. *Eur J Pharm Biopharm* 78 (2), S. 184–188.

Jiang, Cizhong; Zhao, Zhongming (2006). *BMC Genomics* 7 (1), S. 316.

Jiang, J.; Ballinger, C. A.; Wu, Y.; Dai, Q.; Cyr, D. M.; Höhfeld, J.; Patterson, C. (2001): CHIP is a U-box-dependent E3 ubiquitin ligase: identification of Hsc70 as a target for ubiquitylation. *J. Biol. Chem.* 276 (46), S. 42938–42944.

Jiang, X.; Sun, Y.; Chen, S.; Roy, K.; Price, B. D. (2006): The FATC Domains of PIKK Proteins Are Functionally Equivalent and Participate in the Tip60-dependent Activation of DNA-PKcs and ATM. *Journal of Biological Chemistry* 281 (23), S. 15741–15746.

Jiang, Yuanyuan; Sirinupong, Nualpun; Brunzelle, Joseph; Yang, Zhe (2011): Crystal structures of histone and p53 methyltransferase SmyD2 reveal a conformational flexibility of the autoinhibitory C-terminal domain. *PLoS ONE* 6 (6), S. e21640.

Jochmann, Ramona; Holz, Patrick; Sticht, Heinrich; Stürzl, Michael (2013): Validation of the reliability of computational O-GlcNAc prediction. *Biochim. Biophys. Acta* 1844 (2), S. 416–421.

Johnson, Colin A.; White, Darren A.; Lavender, Jayne S.; O'Neill, Laura P.; Turner, Bryan M. (2002): Human class I histone deacetylase complexes show enhanced catalytic activity in the presence of ATP and co-immunoprecipitate with the ATP-dependent chaperone protein Hsp70. *J. Biol. Chem.* 277 (11), S. 9590–9597.

Jones, Meredith B.; Tomiya, Noboru; Betenbaugh, Michael J.; Krag, Sharon S. (2010): Analysis and metabolic engineering of lipid-linked oligosaccharides in glycosylation-deficient CHO cells. *Biochem. Biophys. Res. Commun.* 395 (1), S. 36–41.

Joo, M. S.; Lee, C. G.; Koo, J. H.; Kim, S. G. (2013): miR-125b transcriptionally increased by Nrf2 inhibits AhR repressor, which protects kidney from cisplatin-induced injury. *Cell Death Dis* 4, S. e899.

Juhásová, Barbora; Mentel, Marek; Bhatia-Kiššová, Ingrid; Zeman, Igor; Kolarov, Jordan; Forte, Michael; Polčic, Peter (2011): BH3-only protein Bim inhibits activity of antiapoptotic members of Bcl-2 family when expressed in yeast. *FEBS Lett.* 585 (17), S. 2709–2713.

Juhasz, Kata; Lipp, Anna-Maria; Nimmervoll, Benedikt; Sonnleitner, Alois; Hesse, Jan; Haselgruebler, Thomas; Balogi, Zsolt (2013): The complex function of hsp70 in metastatic cancer. *Cancers (Basel)* 6 (1), S. 42–66.

Kaneko, Tomomi; Kibayashi, Kazuhiko (2012): Mild hypothermia facilitates the expression of cold-inducible RNA-binding protein and heat shock protein 70.1 in mouse brain. *Brain Res.* 1466, S. 128–136.

Kaneko, Yoshihiro; Sato, Ryuji; Aoyagi, Hideki (2010): Evaluation of Chinese hamster ovary cell stability during repeated batch culture for large-scale antibody production. *J. Biosci. Bioeng.* 109 (3), S. 274–280.

Kao, F.; Chasin, L.; PUCK, T. T. (1969): Genetics of somatic mammalian cells. X. Complementation analysis of glycine-requiring mutants. *Proc. Natl. Acad. Sci. U.S.A.* 64 (4), S. 1284–1291.

Kapoor-Vazirani, Priya; Kagey, Jacob D.; Vertino, Paula M. (2011): SUV420H2-mediated H4K20 trimethylation enforces RNA polymerase II promoter-proximal pausing by blocking hMOF-dependent H4K16 acetylation. *Mol. Cell. Biol.* 31 (8), S. 1594–1609.

Kaufmann, H.; Mazur, X.; Fussenegger, M.; Bailey, J. E. (1999): Influence of low temperature on productivity, proteome and protein phosphorylation of CHO cells. *Biotechnol. Bioeng.* 63 (5), S. 573–582.

Kaufman, R. J.; Sharp, P. A. (1982): Amplification and expression of sequences cotransfected with a modular dihydrofolate reductase complementary dna gene. *J. Mol. Biol.* 159 (4), S. 601–621.

Kawai, H. (2001): Dual Role of p300 in the Regulation of p53 Stability. *Journal of Biological Chemistry* 276 (49), S. 45928–45932.

Kefalas, P.; Brown, T. R.; Brickell, P. M. (1995): Signalling by the p60c-src family of protein-tyrosine kinases. *Int. J. Biochem. Cell Biol.* 27 (6), S. 551–563.

Keysar, Stephen B.; Fox, Michael H. (2009): Kinetics of CHO A L mutant expression after treatment with gamma radiation, EMS, and asbestos. *Cytometry A* 75 (5), S. 412–419.

Khamlichi, A. A.; Rocca, A.; Cogné, M. (1994): The effect of intron sequences on expression levels of Ig cDNAs. *Gene* 150 (2), S. 387–390.

Khan, Mahmud Tareq Hassan; Mischiati, Carlo; Ather, Arjumand; Ohyama, Tatsuya; Dedachi, Kenichi; Borgatti, Monica *et al.* (2012): Structure-based analysis of the molecular recognitions between HIV-1 TAR-RNA and transcription factor nuclear factor-kappaB (NFkB). *Curr Top Med Chem* 12 (8), S. 814–827.

Khoei, Samideh; Goliaei, Bahram; Neshasteh-Riz, Ali; Deizadji, Abdolkhalegh (2004): The role of heat shock protein 70 in the thermoresistance of prostate cancer cell line spheroids. *FEBS Letters* 561 (1-3), S. 144–148.

Kim, Hyungjin; Tu, Ho-Chou; Ren, Decheng; Takeuchi, Osamu; Jeffers, John R.; Zambetti, Gerard P. *et al.* (2009a): Stepwise Activation of BAX and BAK by tBID, BIM, and PUMA Initiates Mitochondrial Apoptosis. *Molecular Cell* 36 (3), S. 487–499.

Kim, Ki Chan; Oh, Won Jung; Ko, Kwang Ho; Shin, Chan Young; Wells, David G. (2011): Cyclin B1 expression regulated by cytoplasmic polyadenylation element binding protein in astrocytes. *J. Neurosci.* 31 (34), S. 12118–12128.

Kim, Sung Hyun; Lee, Gyun Min (2007a): Down-regulation of lactate dehydrogenase-A by siRNAs for reduced lactic acid formation of Chinese hamster ovary cells producing thrombopoietin. *Appl. Microbiol. Biotechnol.* 74 (1), S. 152–159.

Kim, Sung Hyun; Lee, Gyun Min (2007b): Functional expression of human pyruvate carboxylase for reduced lactic acid formation of Chinese hamster ovary cells (DG44). *Appl Microbiol Biotechnol* 76 (3), S. 659–665.

Kim, Yeon-Gu; Kim, Jee Yon; Lee, Gyun Min (2009b): Effect of XIAP overexpression on sodium butyrate-induced apoptosis in recombinant Chinese hamster ovary cells producing erythropoietin. *Journal of Biotechnology* 144 (4), S. 299–303.

Kim, Yeon-Gu; Kim, Jee Yon; Mohan, Chaya; Lee, Gyun Min (2009c): Effect of Bcl-xL overexpression on apoptosis and autophagy in recombinant Chinese hamster ovary cells under nutrient-deprived condition. *Biotechnol. Bioeng.* 103 (4), S. 757–766.

Kim, Yeon-Gu; Park, Byoungwoo; Ahn, Jung Oh; Jung, Joon-Ki; Lee, Hong Weon; Lee, Eun Gyo (2012): New cell line development for antibody-producing Chinese hamster ovary cells using split green fluorescent protein. *BMC Biotechnol.* 12, S. 24.

Kino, T.; Hurt, D. E.; Ichijo, T.; Nader, N.; Chrousos, G. P. (2010): Noncoding RNA Gas5 Is a Growth Arrest- and Starvation-Associated Repressor of the Glucocorticoid Receptor. *Science Signaling* 3 (107), S. ra8.

Kishore, S.; Khanna, A.; Zhang, Z.; Hui, J.; Balwierz, P. J.; Stefan, M. *et al.* (2010): The snoRNA MBII-52 (SNORD 115) is processed into smaller RNAs and regulates alternative splicing. *Human Molecular Genetics* 19 (7), S. 1153–1164.

Kiss-László, Z.; Henry, Y.; Bachellerie, J. P.; Caizergues-Ferrer, M.; Kiss, T. (1996): Site-specific ribose methylation of preribosomal RNA: a novel function for small nucleolar RNAs. *Cell* 85 (7), S. 1077–1088.

Knudsen, S. (1999): Promoter2.0: for the recognition of PolII promoter sequences. *Bioinformatics* 15 (5), S. 356–361.

Komarova, Elena Yu; Afanasyeva, Elena A.; Bulatova, Marina M.; Cheetham, Michael E.; Margulis, Boris A.; Guzhova, Irina V. (2004): Downstream caspases are novel targets for the antiapoptotic activity of the molecular chaperone hsp70. *Cell Stress Chaperones* 9 (3), S. 265–275.

Kopnin, Pavel B.; Kravchenko, Irina V.; Furalyov, Vladimir A.; Pylev, Lev N.; Kopnin, Boris P. (2004): Cell type-specific effects of asbestos on intracellular ROS levels, DNA oxidation and G1 cell cycle checkpoint. *Oncogene* 23 (54), S. 8834–8840.

Kosugi, Shunichi; Hasebe, Masako; Tomita, Masaru; Yanagawa, Hiroshi (2009): Systematic identification of cell cycle-dependent yeast nucleocytoplasmic shuttling proteins by prediction of composite motifs. *Proc. Natl. Acad. Sci. U.S.A.* 106 (25), S. 10171–10176.

Krampe, Britta; Al-Rubeai, Mohamed (2010): Cell death in mammalian cell culture: molecular mechanisms and cell line engineering strategies. *Cytotechnology* 62 (3), S. 175–188.

Krampe, Britta; Swiderek, Halina; Al-Rubeai, Mohamed (2008): Transcriptome and proteome analysis of antibody-producing mouse myeloma NS0 cells cultivated at different cell densities in perfusion culture. *Biotechnol. Appl. Biochem.* 50 (Pt 3), S. 133–141.

Krogh, A.; Larsson, B.; Heijne, G. von; Sonnhammer, E. L. (2001): Predicting transmembrane protein topology with a hidden Markov model: application to complete genomes. *J. Mol. Biol.* 305 (3), S. 567–580.

Krug, Alexander W.; Schuster, Claudia; Gassner, Birgit; Freudinger, Ruth; Mildenberger, Sigrid; Troppmair, Jakob; Gekle, Michael (2002): Human epidermal growth factor receptor-1 expression renders Chinese hamster ovary cells sensitive to alternative aldosterone signaling. *J. Biol. Chem.* 277 (48), S. 45892–45897.

Kumar, Niraj; Gammell, Patrick; Meleady, Paula; Henry, Michael; Clynes, Martin (2008): Differential protein expression following low temperature culture of suspension CHO-K1 cells. *BMC Biotechnol.* 8, S. 42.

Ku, Sebastian C. Y.; Ng, Daphne T. W.; Yap, Miranda G. S.; Chao, Sheng-Hao (2008): Effects of overexpression of X-box binding protein 1 on recombinant protein production in Chinese hamster ovary and NS0 myeloma cells. *Biotechnol. Bioeng.* 99 (1), S. 155–164.

Kuystermans, Darrin; Al-Rubeai, Mohamed (2009): cMyc increases cell number through uncoupling of cell division from cell size in CHO cells. *BMC Biotechnol* 9 (1), S. 76.

Kwaks, Ted H. J.; Barnett, Phil; Hemrika, Wieger; Siersma, Tjalling; Sewalt, Richard G. A. B.; Satijn, David P. E. *et al.* (2003): Identification of anti-repressor elements that confer high and stable protein production in mammalian cells. *Nat. Biotechnol.* 21 (5), S. 553–558.

Kwon, Mijung; Lee, Soo Jin; Wang, Yarong; Rybak, Yevangelina; Luna, Alex; Reddy, Srilakshmi *et al.* (2013): Filamin A interacting protein 1-like inhibits WNT signaling and MMP expression to suppress cancer cell invasion and metastasis. *Int. J. Cancer.*

La Cour, Tanja; Kiemer, Lars; Mølgaard, Anne; Gupta, Ramneek; Skriver, Karen; Brunak, Søren (2004): Analysis and prediction of leucine-rich nuclear export signals. *Protein Eng. Des. Sel.* 17 (6), S. 527–536.

Lao, M. S.; Toth, D. (1997): Effects of ammonium and lactate on growth and metabolism of a recombinant Chinese hamster ovary cell culture. *Biotechnol. Prog.* 13 (5), S. 688–691.

Le Bihan, T.; Pinto, D.; Figeys, D. (2001): Nanoflow gradient generator coupled with mu-LC-ESI-MS/MS for protein identification. *Anal. Chem.* 73 (6), S. 1307–1315.

Lee, Min Gyu; Wynder, Christopher; Bochar, Daniel A.; Hakimi, Mohamed-Ali; Cooch, Neil; Shiekhattar, Ramin (2006): Functional interplay between histone demethylase and deacetylase enzymes. *Mol. Cell. Biol.* 26 (17), S. 6395–6402.

Lee, Moon Sue; Kim, Kyoung Wook; Kim, Young Hwan; Lee, Gyun Min (2003): Proteome analysis of antibody-expressing CHO cells in response to hyperosmotic pressure. *Biotechnol. Prog.* 19 (6), S. 1734–1741.

Lee, Yih Yean; Wong, Kathy T.K; Tan, Janice; Toh, Poh Choo; Mao, Yanying; Brusic, Vesna; Yap, Miranda G.S (2009): Overexpression of heat shock proteins (HSPs) in CHO cells for extended culture viability and improved recombinant protein production. *Journal of Biotechnology* 143 (1), S. 34–43.

Lee, Yoontae; Jeon, Kipyoung; Lee, Jun-Tae; Kim, Sunyoung; Kim, V. Narry (2002): MicroRNA maturation: stepwise processing and subcellular localization. *EMBO J.* 21 (17), S. 4663–4670.

Lewis, M. J.; Pelham, H. R. (1985): Involvement of ATP in the nuclear and nucleolar functions of the 70 kd heat shock protein. *EMBO J.* 4 (12), S. 3137–3143.

Lewis, Nathan E.; Liu, Xin; Li, Yuxiang; Nagarajan, Harish; Yerganian, George; O'Brien, Edward *et al.* (2013): Genomic landscapes of Chinese hamster ovary cell lines as revealed by the Cricetulus griseus draft genome. *Nat. Biotechnol.* 31 (8), S. 759–765.

Liang, Li; Li, Xianzheng; Zhang, Xiaojing; Lv, Zhenbing; He, Guoyang; Zhao, Wei *et al.* (2013): MicroRNA-137, an HMGA1 target, suppresses colorectal cancer cell invasion and metastasis in mice by directly targeting FMNL2. *Gastroenterology* 144 (3), S. 624-635.e4.

Liang, Xiquan; Zhao, Jenson; Hajivandi, Mahbod; Wu, Rina; Tao, Janet; Amshey, Joseph W.; Pope, R. Marshall (2006): Quantification of membrane and membrane-bound proteins in normal and malignant breast cancer cells isolated from the same patient with primary breast carcinoma. *J. Proteome Res.* 5 (10), S. 2632–2641.

Liang, Xue-Hai; Liu, Qing; Fournier, Maurille J. (2009): Loss of rRNA modifications in the decoding center of the ribosome impairs translation and strongly delays pre-rRNA processing. *RNA* 15 (9), S. 1716–1728.

Liew, Jane C.J; Tan, Wen Siang; Alitheen, Noorjahan Banu Mohamed; Chan, Eng-Seng; Tey, Beng Ti (2010): Over-expression of the X-linked inhibitor of apoptosis protein (XIAP) delays serum deprivation-induced apoptosis in CHO-K1 cells. *Journal of Bioscience and Bioengineering* 110 (3), S. 338–344.

Li, Haiyan; Wang, Qun; Gao, Fei; Zhu, Faliang; Wang, Xiaoyan; Zhou, Chengjun *et al.* (2008): Reduced expression of PDCD5 is associated with high-grade astrocytic gliomas. *Oncol. Rep.* 20 (3), S. 573–579.

Li, Hangyu; Li, Yan; Liu, Dan; Sun, Hongzhi; Su, Dongming; Yang, Fuquan; Liu, Jingang (2013a): Extracellular HSP70/HSP70-PCs promote epithelial-mesenchymal transition of hepatocarcinoma cells. *PLoS ONE* 8 (12), S. e84759.

Li, Jincai; Wong, Chun Loong; Vijayasankaran, Natarajan; Hudson, Terry; Amanullah, Ashraf (2012): Feeding lactate for CHO cell culture processes: impact on culture metabolism and performance. *Biotechnol. Bioeng.* 109 (5), S. 1173–1186.

Li, Jun-Jie; Pei, Yan; Zhou, Guang-Bin; Suo, Lun; Wang, Yan-Ping; Wu, Guo-Quan *et al.* (2011): Histone deacetyltransferase1 expression in mouse oocyte and their in vitro-fertilized embryo: effect of oocyte vitrification. *Cryo Letters* 32 (1), S. 13–20.

Lim, Sing Fee; Chuan, Kok Hwee; Liu, Sen; Loh, Sophia O. H.; Chung, Beatrice Y. F.; Ong, Chin Chew; Song, Zhiwei (2006): RNAi suppression of Bax and Bak enhances viability in fed-batch cultures of CHO cells. *Metab. Eng.* 8 (6), S. 509–522.

Lin, Jye-Yee; Ohshima, Takayuki; Shimotohno, Kunitada (2004): Association of Ubc9, an E2 ligase for SUMO conjugation, with p53 is regulated by phosphorylation of p53. *FEBS Lett.* 573 (1-3), S. 15–18.

Li, Rong; Wei, Jie; Jiang, Cong; Liu, Dongmei; Deng, Lu; Zhang, Kai; Wang, Ping (2013b): Akt SUMOylation regulates cell proliferation and tumorigenesis. *Cancer Res.* 73 (18), S. 5742–5753.

Liste-Calleja, Leticia; Lecina, Martí; Cairó, Jordi Joan (2013): HEK293 cell culture media study towards bioprocess optimization: Animal derived component free and animal derived component containing platforms. *Journal of Bioscience and Bioengineering.*

Litvinov, Ivan V.; Kupper, Thomas S.; Sasseville, Denis (2012): The role of AHI1 and CDKN1C in cutaneous T-cell lymphoma progression. *Exp. Dermatol.* 21 (12), S. 964–966.

Liu, Chi-Hsien; Chen, Li-Hsin (2007): Promotion of recombinant macrophage colony stimulating factor production by dimethyl sulfoxide addition in Chinese hamster ovary cells. *J. Biosci. Bioeng.* 103 (1), S. 45–49.

Liu, Lin; Li, Yinhu; Li, Siliang; Hu, Ni; He, Yimin; Pong, Ray *et al.* (2012): Comparison of Next-Generation Sequencing Systems. *Journal of Biomedicine and Biotechnology* 2012 (7), S. 1–11.

Liu, Qinghuai; Gao, Juanyu; Chen, Xi; Chen, Yuxin; Chen, Jie; Wang, Saiqun *et al.* (2008): HBP21: A Novel Member of TPR Motif Family, as a Potential Chaperone of Heat Shock Protein 70 in Proliferative Vitreoretinopathy (PVR) and Breast Cancer. *Mol Biotechnol* 40 (3), S. 231–240.

Liu, Xianhong; Daskal, Ierachmiel; Kwok, Simon C. M. (2003): Effects of PTX1 expression on growth and tumorigenicity of the prostate cancer cell line PC-3. *DNA Cell Biol.* 22 (7), S. 469–474.

Liu, Xiao-Ming; Yang, Fei-Fei; Yuan, Yi-Feng; Zhai, Rui; Huo, Li-Jun (2013a): SUMOylation of mouse p53b by SUMO-1 promotes its pro-apoptotic function in ovarian granulosa cells. *PLoS ONE* 8 (5), S. e63680.

Liu, Yuting; Hu, Wenchao; Murakawa, Yasuhiro; Yin, Jingwen; Wang, Gang; Landthaler, Markus; Yan, Jun (2013b): Cold-induced RNA-binding proteins regulate circadian gene expression by controlling alternative polyadenylation. *Sci. Rep.* 3.

Liu, Zhihong; Wang, Wei; Jiang, Juntao; Bao, Erdun; Xu, Dongliang; Zeng, Yigang *et al.* (2013c): Downregulation of GAS5 promotes bladder cancer cell proliferation, partly by regulating CDK6. *PLoS ONE* 8 (9), S. e73991.

Luo, Chonglin; Tetteh, Paul W.; Merz, Patrick R.; Dickes, Elke; Abukiwan, Alia; Hotz-Wagenblatt, Agnes *et al.* (2013a): miR-137 inhibits the invasion of melanoma cells through downregulation of multiple oncogenic target genes. *J. Invest. Dermatol.* 133 (3), S. 768–775.

Luo, D.; Bu, Y.; Ma, J.; Rajput, S.; He, Y.; Cai, G. *et al.* (2013b): Heat Shock Protein 90- Mediates Aldo-Keto Reductase 1B10 (AKR1B10) Protein Secretion through Secretory Lysosomes. *Journal of Biological Chemistry* 288 (51), S. 36733–36740.

Lu, Xiongxiong; Fang, Yuan; Wang, Zhengting; Xie, Junjie; Zhan, Qian; Deng, Xiaxing *et al.* (2013): Downregulation of gas5 increases pancreatic cancer cell proliferation by regulating CDK6. *Cell Tissue Res.* 354 (3), S. 891–896.

Lv, Zhenbing; Zou, Huichun; Peng, Kaiwen; Wang, Jianmei; Ding, Yi; Li, Yuling *et al.* (2013): The Suppressive Role and Aberrant Promoter Methylation of BTG3 in the Progression of Hepatocellular Carcinoma. *PLoS ONE* 8 (10), S. e77473.

Lynch, J. T.; Somerville, T. D. D.; Spencer, G. J.; Huang, X.; Somervaille, T. C. P. (2013): TTC5 is required to prevent apoptosis of acute myeloid leukemia stem cells. *Cell Death Dis* 4, S. e573.

Lyu, Jungmook; Jho, Eek-Hoon; Lu, Wange (2011): Smek promotes histone deacetylation to suppress transcription of Wnt target gene brachyury in pluripotent embryonic stem cells. *Cell Res.* 21 (6), S. 911–921.

Maeda, Y.; Tanaka, S.; Hino, J.; Kangawa, K.; Kinoshita, T. (2000): Human dolichol-phosphate-mannose synthase consists of three subunits, DPM1, DPM2 and DPM3. *EMBO J.* 19 (11), S. 2475–2482.

Maeda, Y.; Tomita, S.; Watanabe, R.; Ohishi, K.; Kinoshita, T. (1998): DPM2 regulates biosynthesis of dolichol phosphate-mannose in mammalian cells: correct subcellular localization and stabilization of DPM1, and binding of dolichol phosphate. *EMBO J.* 17 (17), S. 4920–4929.

Maerki, Sarah; Olma, Michael H.; Staubli, Titu; Steigemann, Patrick; Gerlich, Daniel W.; Quadroni, Manfredo *et al.* (2009): The Cul3-KLHL21 E3 ubiquitin ligase targets aurora B to midzone microtubules in anaphase and is required for cytokinesis. *J. Cell Biol.* 187 (6), S. 791–800.

Magin, Simon; Saha, Janapriya; Wang, Minli; Mladenova, Veronika; Coym, Nadine; Iliakis, George (2013): Lipofection and nucleofection of substrate plasmid can generate widely different readings of DNA end-joining efficiency in different cell lines. *DNA Repair (Amst.)* 12 (2), S. 148–160.

Majid, Shahana; Dar, Altaf A.; Ahmad, Ardalan E.; Hirata, Hiroshi; Kawakami, Kazumori; Shahryari, Varahram *et al.* (2009): BTG3 tumor suppressor gene promoter demethylation, histone modification and cell cycle arrest by genistein in renal cancer. *Carcinogenesis* 30 (4), S. 662–670.

Majors, Brian S.; Arden, Nilou; Oyler, George A.; Chiang, Gisela G.; Pederson, Nels E.; Betenbaugh, Michael J. (2008): E2F-1 overexpression increases viable cell density in batch cultures of Chinese hamster ovary cells. *Journal of Biotechnology* 138 (3-4), S. 103–106.

Majors, Brian S.; Betenbaugh, Michael J.; Pederson, Nels E.; Chiang, Gisela G. (2009): Mcl-1 overexpression leads to higher viabilities and increased production of humanized monoclonal antibody in Chinese hamster ovary cells. *Biotechnol. Prog.* 25 (4), S. 1161–1168.

Majors, Brian S.; Chiang, Gisela G.; Pederson, Nels E.; Betenbaugh, Michael J. (2012): Directed evolution of mammalian anti-apoptosis proteins by somatic hypermutation. *Protein Eng. Des. Sel.* 25 (1), S. 27–38.

Makarova, Julia A.; Kramerov, Dmitri A. (2009): Analysis of C/D box snoRNA genes in vertebrates: The number of copies decreases in placental mammals. *Genomics* 94 (1), S. 11–19.

Mambula, Salamatu S.; Stevenson, Mary Ann; Ogawa, Kishiko; Calderwood, Stuart K. (2007): Mechanisms for Hsp70 secretion: Crossing membranes without a leader. *Methods* 43 (3), S. 168–175.

Manzoor, Rashid; Kuroda, Kazumichi; Yoshida, Reiko; Tsuda, Yoshimi; Fujikura, Daisuke; Miyamoto, Hiroko *et al.* (2014): Heat Shock Protein 70 Modulates Influenza A Virus Polymerase Activity. *J. Biol. Chem.*

Mao, H. (2004): Hsp72 Interacts with Paxillin and Facilitates the Reassembly of Focal Adhesions during Recovery from ATP Depletion. *Journal of Biological Chemistry* 279 (15), S. 15472–15480.

Mao, Haiping; Li, Fanghong; Ruchalski, Kathleen; Mosser, Dick D.; Schwartz, John H.; Wang, Yihan; Borkan, Steven C. (2003): hsp72 inhibits focal adhesion kinase degradation in ATP-depleted renal epithelial cells. *J. Biol. Chem.* 278 (20), S. 18214–18220.

Mao, R-F; Rubio, V.; Chen, H.; Bai, L.; Mansour, O. C.; Shi, Z-Z (2013): OLA1 protects cells in heat shock by stabilizing HSP70. *Cell Death Dis* 4 (2), S. e491.

Maragkakis, M.; Reczko, M.; Simossis, V. A.; Alexiou, P.; Papadopoulos, G. L.; Dalamagas, T. *et al.* (2009): DIANA-microT web server: elucidating microRNA functions through target prediction. *Nucleic Acids Research* 37 (Web Server), S. W273.

Marban, Céline; Suzanne, Stella; Dequiedt, Franck; Walque, Stéphane de; Redel, Laetitia; van Lint, Carine *et al.* (2007): Recruitment of chromatin-modifying enzymes by CTIP2 promotes HIV-1 transcriptional silencing. *EMBO J.* 26 (2), S. 412–423.

Marchler-Bauer, Aron; Zheng, Chanjuan; Chitsaz, Farideh; Derbyshire, Myra K.; Geer, Lewis Y.; Geer, Renata C. *et al.* (2013): CDD: conserved domains and protein three-dimensional structure. *Nucleic Acids Res.* 41 (Database issue), S. D348-52.

Marinova, Zoya; Leng, Yan; Leeds, Peter; Chuang, De-Maw (2011): Histone deacetylase inhibition alters histone methylation associated with heat shock protein 70 promoter modifications in astrocytes and neurons. *Neuropharmacology* 60 (7-8), S. 1109–1115.

Marschalek, A.; Finke, S.; Schwemmle, M.; Mayer, D.; Heimrich, B.; Stitz, L.; Conzelmann, K.-K (2009): Attenuation of Rabies Virus Replication and Virulence by Picornavirus Internal Ribosome Entry Site Elements. *Journal of Virology* 83 (4), S. 1911–1919.

Matsumura, Yoshihiro; Sakai, Juro; Skach, William R. (2013): Endoplasmic reticulum protein quality control is determined by cooperative interactions between Hsp/c70 protein and the CHIP E3 ligase. *J. Biol. Chem.* 288 (43), S. 31069–31079.

Ma, Zhiyuan; Vocadlo, David J.; Vosseller, Keith (2013): Hyper-O-GlcNAcylation is anti-apoptotic and maintains constitutive NF-κB activity in pancreatic cancer cells. *J. Biol. Chem.* 288 (21), S. 15121–15130.

Meleady, Paula; Doolan, Padraig; Henry, Michael; Barron, Niall; Keenan, Joanne; O'Sullivan, Finbar *et al.* (2011): Sustained productivity in recombinant Chinese hamster ovary (CHO) cell lines: proteome analysis of the molecular basis for a process-related phenotype. *BMC Biotechnol.* 11, S. 78.

Melville, Mark; Doolan, Padraig; Mounts, William; Barron, Niall; Hann, Louane; Leonard, Mark *et al.* (2011): Development and characterization of a Chinese hamster ovary cell-specific oligonucleotide microarray. *Biotechnol. Lett.* 33 (9), S. 1773–1779.

Misaghi, Shahram; Qu, Yan; Snowden, Andrew; Chang, Jennifer; Snedecor, Brad (2013): Resilient immortals, characterizing and utilizing bax/bak deficient chinese hamster ovary (CHO) cells for high titer antibody production. *Biotechnol. Prog.*

Misawa, Kazuharu; Kikuno, Reiko F. (2009): Evaluation of the effect of CpG hypermutability on human codon substitution. *Gene* 431 (1-2), S. 18–22.

Mochizuki, Yasuhiro; Ohashi, Riuko; Kawamura, Takeshi; Iwanari, Hiroko; Kodama, Tatsuhiko; Naito, Makoto; Hamakubo, Takao (2013): Phosphatidylinositol 3-phosphatase myotubularin-related protein 6 (MTMR6) is regulated by small GTPase Rab1B in the early secretory and autophagic pathways. *J. Biol. Chem.* 288 (2), S. 1009–1021.

Mohan, Chaya; Lee, Gyun Min (2010): Effect of inducible co-overexpression of protein disulfide isomerase and endoplasmic reticulum oxidoreductase on the specific antibody productivity of recombinant Chinese hamster ovary cells. *Biotechnol. Bioeng.* 107 (2), S. 337–346.

Moncman, C. L.; Wang, K. (1995): Nebulette: a 107 kD nebulin-like protein in cardiac muscle. *Cell Motil. Cytoskeleton* 32 (3), S. 205–225.

Moran, M. F.; Polakis, P.; McCormick, F.; Pawson, T.; Ellis, C. (1991): Protein-tyrosine kinases regulate the phosphorylation, protein interactions, subcellular distribution, and activity of p21ras GTPase-activating protein. *Mol. Cell. Biol.* 11 (4), S. 1804–1812.

Mourtada-Maarabouni, M.; Hasan, A. M.; Farzaneh, F.; Williams, G. T. (2010): Inhibition of Human T-Cell Proliferation by Mammalian Target of Rapamycin (mTOR) Antagonists Requires Noncoding RNA Growth-Arrest-Specific Transcript 5 (GAS5). *Molecular Pharmacology* 78 (1), S. 19–28.

Mourtada-Maarabouni, Mirna; Williams, Gwyn T. (2013): Growth Arrest on Inhibition of Nonsense-Mediated Decay Is Mediated by Noncoding RNA GAS5. *Biomed Res Int* 2013, S. 358015.

Nakagawa, Hiroyuki; Suzuki, Hiroshi; Machida, Satoshi; Suzuki, Junko; Ohashi, Kazuyo; Jin, Mingyue *et al.* (2009): Contribution of the LIM domain and nebulin-repeats to the interaction of Lasp-2 with actin filaments and focal adhesions. *PLoS ONE* 4 (10), S. e7530.

Natoli, Thomas A.; Gareski, Tiffany C.; Dackowski, William R.; Smith, Laurie; Bukanov, Nikolay O.; Russo, Ryan J. *et al.* (2008): Pkd1 and Nek8 mutations affect cell-cell adhesion and cilia in cysts formed in kidney organ cultures. *Am. J. Physiol. Renal Physiol.* 294 (1), S. F73-83.

Názer, Ezequiel; Verdún, Ramiro E.; Sánchez, Daniel O. (2012): Severe heat shock induces nucleolar accumulation of mRNAs in Trypanosoma cruzi. *PLoS ONE* 7 (8), S. e43715.

Neudecker, Sabine; Walz, Rebecca; Menon, Kiran; Maier, Elena; Bihoreau, Marie-Therese; Obermüller, Nicholas *et al.* (2010): Transgenic Overexpression of Anks6(p.R823W) Causes Polycystic Kidney Disease in Rats. *The American Journal of Pathology* 177 (6), S. 3000–3009.

Neutelings, Thibaut; Lambert, Charles A.; Nusgens, Betty V.; Colige, Alain C. (2013): Effects of mild cold shock (25°C) followed by warming up at 37 °C on the cellular stress response. *PLoS ONE* 8 (7), S. e69687.

Nishiyama, H.; Itoh, K.; Kaneko, Y.; Kishishita, M.; Yoshida, O.; Fujita, J. (1997): A glycine-rich RNA-binding protein mediating cold-inducible suppression of mammalian cell growth. *J. Cell Biol.* 137 (4), S. 899–908.

Nivitchanyong, Toey; Martinez, Amanda; Ishaque, Adiba; Murphy, John E.; Konstantinov, Konstantin; Betenbaugh, Michael J.; Thrift, John (2007): Anti-apoptotic genes Aven and E1B-19K enhance performance of BHK cells engineered to express recombinant factor VIII in batch and low perfusion cell culture. *Biotechnol. Bioeng.* 98 (4), S. 825–841.

Nobes, C. D.; Hall, A. (1995): Rho, rac, and cdc42 GTPases regulate the assembly of multimolecular focal complexes associated with actin stress fibers, lamellipodia, and filopodia. *Cell* 81 (1), S. 53–62.

Nordgård, Oddmund; Kvaløy, Jan Terje; Farmen, Ragne Kristin; Heikkilä, Reino (2006): Error propagation in relative real-time reverse transcription polymerase chain reaction quantification models: the balance between accuracy and precision. *Anal. Biochem.* 356 (2), S. 182–193.

Okitsu, C. Y.; Hsieh, C.-L (2007): DNA Methylation Dictates Histone H3K4 Methylation. *Molecular and Cellular Biology* 27 (7), S. 2746–2757.

Okitsu, C. Y.; Hsieh, J. C. F.; Hsieh, C. L. (2010): Transcriptional Activity Affects the H3K4me3 Level and Distribution in the Coding Region. *Molecular and Cellular Biology* 30 (12), S. 2933–2946.

O'Neill, J. P.; Fuscoe, J. C.; Hsie, A. W. (1978): Mutagenicity of heterocyclic nitrogen mustards (ICR compounds) in cultured mammalian cells. *Cancer Res.* 38 (3), S. 506–509.

Ørom, Ulf Andersson; Nielsen, Finn Cilius; Lund, Anders H. (2008): MicroRNA-10a Binds the 5'UTR of Ribosomal Protein mRNAs and Enhances Their Translation. *Molecular Cell* 30 (4), S. 460–471.

Osterlehner, Andrea; Simmeth, Silke; Göpfert, Ulrich (2011): Promoter methylation and transgene copy numbers predict unstable protein production in recombinant Chinese hamster ovary cell lines. *Biotechnol. Bioeng.* 108 (11), S. 2670–2681.

Ozcan, Sabire; Andrali, Sreenath S.; Cantrell, Jamie E. L. (2010): Modulation of transcription factor function by O-GlcNAc modification. *Biochim. Biophys. Acta* 1799 (5-6), S. 353–364.

Özgür, Emre; Mert, Ufuk; Isin, Mustafa; Okutan, Murat; Dalay, Nejat; Gezer, Ugur (2013): Differential expression of long non-coding RNAs during genotoxic stress-induced apoptosis in HeLa and MCF-7 cells. *Clin. Exp. Med.* 13 (2), S. 119–126.

Pagni, M. (2001): trEST, trGEN and Hits: access to databases of predicted protein sequences. *Nucleic Acids Research* 29 (1), S. 148–151.

Panaviene, Zivile; Moncman, Carole L. (2007): Linker region of nebulin family members plays an important role in targeting these molecules to cellular structures. *Cell Tissue Res.* 327 (2), S. 353–369.

Pannekoek, Willem-Jan; Kooistra, Matthijs R. H.; Zwartkruis, Fried J. T.; Bos, Johannes L. (2009): Cell-cell junction formation: the role of Rap1 and Rap1 guanine nucleotide exchange factors. *Biochim. Biophys. Acta* 1788 (4), S. 790–796.

Pappas, Christopher T.; Bliss, Katherine T.; Zieseniss, Anke; Gregorio, Carol C. (2011): The Nebulin family: an actin support group. *Trends in Cell Biology* 21 (1), S. 29–37.

Park, Jin-Ah; Kim, Ae-Jin; Kang, Yoonsung; Jung, Yu-Jin; Kim, Hyong Kyu; Kim, Keun-Cheol (2011): Deacetylation and methylation at histone H3 lysine 9 (H3K9) coordinate chromosome condensation during cell cycle progression. *Mol. Cells* 31 (4), S. 343–349.

Paul, I.; Ahmed, S. F.; Bhowmik, A.; Deb, S.; Ghosh, M. K. (2012): The ubiquitin ligase CHIP regulates c-Myc stability and transcriptional activity. *Oncogene* 32 (10), S. 1284–1295.

Paulin, R. P.; Ho, T.; Balzer, H. J.; Holliday, R. (1998): Gene silencing by DNA methylation and dual inheritance in Chinese hamster ovary cells. *Genetics* 149 (2), S. 1081–1088.

Pegoraro, Silvia; Ros, Gloria; Piazza, Silvano; Sommaggio, Roberta; Ciani, Yari; Rosato, Antonio *et al.* (2013): HMGA1 promotes metastatic processes in basal-like breast cancer regulating EMT and stemness. *Oncotarget* 4 (8), S. 1293–1308.

Pekowska, Aleksandra; Benoukraf, Touati; Zacarias-Cabeza, Joaquin; Belhocine, Mohamed; Koch, Frederic; Holota, Hélène *et al.* (2011): H3K4 tri-methylation provides an epigenetic signature of active enhancers. *EMBO J.* 30 (20), S. 4198–4210.

Perez, M. L.; Stamato, T. D. (1999): Time versus replication dependence of EMS-induced delayed mutation in Chinese hamster cells. *Mutat. Res.* 423 (1-2), S. 55–63.

Perrot, Gwenn; Langlois, Benoit; Devy, Jérôme; Jeanne, Albin; Verzeaux, Laurie; Almagro, Sébastien *et al.* (2012): LRP-1--CD44, a new cell surface complex regulating tumor cell adhesion. *Mol. Cell. Biol.* 32 (16), S. 3293–3307.

Petersen, Bent; Lundegaard, Claus; Petersen, Thomas Nordahl (2010): NetTurnP--neural network prediction of beta-turns by use of evolutionary information and predicted protein sequence features. *PLoS ONE* 5 (11), S. e15079.

Petersen, Bent; Petersen, Thomas Nordahl; Andersen, Pernille; Nielsen, Morten; Lundegaard, Claus (2009): A generic method for assignment of reliability scores applied to solvent accessibility predictions. *BMC Struct. Biol.* 9, S. 51.

Petersen, Thomas Nordahl; Brunak, Søren; Heijne, Gunnar von; Nielsen, Henrik (2011): SignalP 4.0: discriminating signal peptides from transmembrane regions. *Nat. Methods* 8 (10), S. 785–786.

Petters, Edyta; Krowarsch, Daniel; Otlewski, Jacek (2013): Design, expression and characterization of a highly stable tetratricopeptide-based protein scaffold for phage display application. *Acta Biochim. Pol.* 60 (4), S. 585–590.

Pickard, M. R.; Mourtada-Maarabouni, M.; Williams, G. T. (2013): Long non-coding RNA GAS5 regulates apoptosis in prostate cancer cell lines. *Biochim. Biophys. Acta* 1832 (10), S. 1613–1623.

Pierleoni, Andrea; Martelli, Pier; Casadio, Rita (2008): PredGPI: a GPI-anchor predictor. *BMC Bioinformatics* 9 (1), S. 392.

Pilbrough, Warren; Munro, Trent P.; Gray, Peter (2009): Intraclonal protein expression heterogeneity in recombinant CHO cells. *PLoS ONE* 4 (12), S. e8432.

Pilotte, Julie; Dupont-Versteegden, Esther E.; Vanderklish, Peter W. (2011): Widespread regulation of miRNA biogenesis at the Dicer step by the cold-inducible RNA-binding protein, RBM3. *PLoS ONE* 6 (12), S. e28446.

Plötz, Michael; Gillissen, Bernhard; Quast, Sandra-Annika; Berger, Anja; Daniel, Peter T.; Eberle, Jürgen (2013): The BH3-only protein BimL overrides Bcl-2-mediated apoptosis resistance in melanoma cells. *Cancer Lett.*

Prentice, Holly L.; Ehrenfels, Barbara N.; Sisk, William P. (2007): Improving performance of mammalian cells in fed-batch processes through "bioreactor evolution". *Biotechnol. Prog.* 23 (2), S. 458–464.

PUCK, T. T.; CIECIURA, S. J.; ROBINSON, A. (1958): Genetics of somatic mammalian cells. III. Long-term cultivation of euploid cells from human and animal subjects. *J. Exp. Med.* 108 (6), S. 945–956.

Puig, Oscar; Bragado-Nilsson, Elisabeth; Koski, Terhi; Séraphin, Bertrand (2007): The U1 snRNP-associated factor Luc7p affects 5' splice site selection in yeast and human. *Nucleic Acids Res.* 35 (17), S. 5874–5885.

Qiao, Hui-Ping; Gao, Wei-Shi; Huo, Jian-Xin; Yang, Zhan-Shan (2013): Long non-coding RNA GAS5 functions as a tumor suppressor in renal cell carcinoma. *Asian Pac. J. Cancer Prev.* 14 (2), S. 1077–1082.

Qi, Q.; Liu, X.; Brat, D. J.; Ye, K. (2013): Merlin sumoylation is required for its tumor suppressor activity. *Oncogene.*

Qu, Li; Ju, Ji Yu; Chen, Shuang Ling; Shi, Yan; Xiang, Zhi Guang; Zhou, Yi Qun et al. (2006): Inhibition of the α-mannosidase Man2c1 gene expression enhances adhesion of Jurkat cells. *Cell Res* 16 (7), S. 622–631.

Radhakrishnan, Prakash; Basma, Hesham; Klinkebiel, David; Christman, Judith; Cheng, Pi-Wan (2008): Cell type-specific activation of the cytomegalovirus promoter by dimethylsulfoxide and 5-aza-2'-deoxycytidine. *Int. J. Biochem. Cell Biol.* 40 (9), S. 1944–1955.

Rai, S. P.; Luthra, R.; Kumar, S. (2003): Salt-tolerant mutants in glycophytic salinity response (GSR) genes in Catharanthus roseus. *Theor. Appl. Genet.* 106 (2), S. 221–230.

Ren, J.; Wen, L.; Gao, X.; Jin, C.; Xue, Y.; Yao, X. (2008): CSS-Palm 2.0: an updated software for palmitoylation sites prediction. *Protein Engineering Design and Selection* 21 (11), S. 639–644.

Ren, Jian; Gao, Xinjiao; Jin, Changjiang; Zhu, Mei; Wang, Xiwei; Shaw, Andrew et al. (2009): Systematic study of protein sumoylation: Development of a site-specific predictor of SUMOsp 2.0. *Proteomics* 9 (12), S. 3409–3412.

Ribeiro, Orquídea; Magalhães, Frederico; Aguiar, Tatiana Q.; Wiebe, Marilyn G.; Penttilä, Merja; Domingues, Lucília (2013): Random and direct mutagenesis to enhance protein secretion in Ashbya gossypii. *bioe* 4 (5), S. 322–331.

Riesco, Marta F.; Robles, Vanesa; Orban, Laszlo (2013): Cryopreservation Causes Genetic and Epigenetic Changes in Zebrafish Genital Ridges. *PLoS ONE* 8 (6), S. e67614.

Ringold, G.; Dieckmann, B., Lee, F. (1981): Co-expression and amplification of dihydrofolate reductase cDNA and the Escherichia coli XGPRT gene in Chinese hamster ovary cells. *J. Mol. Appl. Genet.* 1 (3), S. 165–175.

Roberts, Jonathan R.; Rowe, Peter A.; Demaine, Andrew G. (2002): Activation of NF-kappaB and MAP kinase cascades by hypothermic stress in endothelial cells. *Cryobiology* 44 (2), S. 161–169.

Rogers, J.; Wall, R. (1980): A mechanism for RNA splicing. *Proc. Natl. Acad. Sci. U.S.A.* 77 (4), S. 1877–1879.

Roof, R. W.; Haskell, M. D.; Dukes, B. D.; Sherman, N.; Kinter, M.; Parsons, S. J. (1998): Phosphotyrosine (p-Tyr)-dependent and -independent mechanisms of p190 RhoGAP-p120 RasGAP interaction: Tyr 1105 of p190, a substrate for c-Src, is the sole p-Tyr mediator of complex formation. *Mol. Cell. Biol.* 18 (12), S. 7052–7063.

Roskoski, Robert (2004): Src protein–tyrosine kinase structure and regulation. *Biochemical and Biophysical Research Communications* 324 (4), S. 1155–1164.

Sakurai, Toshiharu; Itoh, Katsuhiko; Liu, Yu; Higashitsuji, Hiroaki; Sumitomo, Yasuhiko; Sakamaki, Kazuhiro; Fujita, Jun (2005): Low temperature protects mammalian cells from apoptosis initiated by various stimuli in vitro. *Exp. Cell Res.* 309 (2), S. 264–272.

Salicioni, A. M. (2003): Low Density Lipoprotein Receptor-related Protein-1 Promotes 1 Integrin Maturation and Transport to the Cell Surface. *Journal of Biological Chemistry* 279 (11), S. 10005–10012.

Sambrook, Joseph; Fritsch, Edward F.; Maniatis, Tom (2001): Molecular cloning. A laboratory manual. 3. Aufl. Cold Spring Harbor, NY: Cold Spring Harbor Laboratory Press.

Santiago, Aleixo; Li, Dawei; Zhao, Lisa Y.; Godsey, Adam; Liao, Daiqing (2013): p53 SUMOylation promotes its nuclear export by facilitating its release from the nuclear export receptor CRM1. *Mol. Biol. Cell* 24 (17), S. 2739–2752.

Saraste, Matti; Sibbald, Peter R.; Wittinghofer, Alfred (1990): The P-loop — a common motif in ATP- and GTP-binding proteins. *Trends in Biochemical Sciences* 15 (11), S. 430–434.

Sauerwald, Tina M.; Betenbaugh, Michael J.; Oyler, George A. (2002): Inhibiting apoptosis in mammalian cell culture using the caspase inhibitor XIAP and deletion mutants. *Biotechnol. Bioeng.* 77 (6), S. 704–716.

Saxena, Tanvi; Tandon, Bhavna; Sharma, Shivani; Chameettachal, Shibu; Ray, Pratima; Ray, Alok R.; Kulshreshtha, Ritu (2013): Combined miRNA and mRNA signature identifies key molecular players and pathways involved in chikungunya virus infection in human cells. *PLoS ONE* 8 (11), S. e79886.

Schefe, Jan H.; Lehmann, Kerstin E.; Buschmann, Ivo R.; Unger, Thomas; Funke-Kaiser, Heiko (2006): Quantitative real-time RT-PCR data analysis: current concepts and the novel "gene expression's CT difference" formula. *J. Mol. Med.* 84 (11), S. 901–910.

Scheufler, C.; Brinker, A.; Bourenkov, G.; Pegoraro, S.; Moroder, L.; Bartunik, H. *et al.* (2000): Structure of TPR domain-peptide complexes: critical elements in the assembly of the Hsp70-Hsp90 multichaperone machine. *Cell* 101 (2), S. 199–210.

Segal, J. A.; Barnett, J. L.; Crawford, D. L. (1999): Functional analyses of natural variation in Sp1 binding sites of a TATA-less promoter. *J. Mol. Evol.* 49 (6), S. 736–749.

Shah, Sandeep N.; Cope, Leslie; Poh, Weijie; Belton, Amy; Roy, Sujayita; Talbot, C. Conover *et al.* (2013): HMGA1: a master regulator of tumor progression in triple-negative breast cancer cells. *PLoS ONE* 8 (5), S. e63419.

Shah, Sandeep N.; Resar, Linda M. S. (2012): High mobility group A1 and cancer: potential biomarker and therapeutic target. *Histol. Histopathol.* 27 (5), S. 567–579.

Shajahan, Ayesha N.; Riggins, Rebecca B.; Clarke, Robert (2009): The role of X-box binding protein-1 in tumorigenicity. *Drug News Perspect.* 22 (5), S. 241–246.

Shao, Peng; Yang, Jian-Hua; Zhou, Hui; Guan, Dao-Gang; Qu, Liang-Hu (2009): Genome-wide analysis of chicken snoRNAs provides unique implications for the evolution of vertebrate snoRNAs. *BMC Genomics* 10, S. 86.

Sharma, S. V. (1998): Rapid recruitment of p120RasGAP and its associated protein, p190RhoGAP, to the cytoskeleton during integrin mediated cell-substrate interaction. *Oncogene* 17 (3), S. 271–281.

Shevchenko, A.; Sunyaev, S.; Loboda, A.; Bork, P.; Ens, W.; Standing, K. G. (2001): Charting the proteomes of organisms with unsequenced genomes by MALDI-quadrupole time-of-flight mass spectrometry and BLAST homology searching. *Anal. Chem.* 73 (9), S. 1917–1926.

Shin, Mi Ra; Choi, Hye Won; Kim, Myo Kyung; Lee, Sun Hee; Lee, Hyoung-Song; Lim, Chun Kyu (2011): In vitro development and gene expression of frozen-thawed 8-cell stage mouse embryos following slow freezing or vitrification. *Clin Exp Reprod Med* 38 (4), S. 203.

Shi, Xuefei; Sun, Ming; Liu, Hongbing; Yao, Yanwen; Kong, Rong; Chen, Fangfang; Song, Yong (2013): A critical role for the long non-coding RNA GAS5 in proliferation and apoptosis in non-small-cell lung cancer. *Mol. Carcinog.*

Shrestha, Smeeta; Cao, Shenglan; Lin, Valerie C. L. (2012): The local microenvironment instigates the regulation of mammary tetratricopeptide repeat domain 9A during lactation and involution through local regulation of the activity of estrogen receptor α. *Biochem. Biophys. Res. Commun.* 426 (1), S. 65–70.

Sidera, Katerina; Gaitanou, Maria; Stellas, Dimitris; Matsas, Rebecca; Patsavoudi, Evangelia (2008): A critical role for HSP90 in cancer cell invasion involves interaction with the extracellular domain of HER-2. *J. Biol. Chem.* 283 (4), S. 2031–2041.

Sigrist, C. J. A.; Cerutti, L.; Castro, E. de; Langendijk-Genevaux, P. S.; Bulliard, V.; Bairoch, A.; Hulo, N. (2009): PROSITE, a protein domain database for functional characterization and annotation. *Nucleic Acids Research* 38 (Database), S. D161.

Singh, Ravi K.; Cooper, Thomas A. (2012): Pre-mRNA splicing in disease and therapeutics. *Trends in Molecular Medicine* 18 (8), S. 472–482.

Slikker, W.; Desai, V. G.; Duhart, H.; Feuers, R.; Imam, S. Z. (2001): Hypothermia enhances bcl-2 expression and protects against oxidative stress-induced cell death in Chinese hamster ovary cells. *Free Radic. Biol. Med.* 31 (3), S. 405–411.

Smith, C. M.; Steitz, J. A. (1998): Classification of gas5 as a multi-small-nucleolar-RNA (snoRNA) host gene and a member of the 5'-terminal oligopyrimidine gene family reveals common features of snoRNA host genes. *Mol. Cell. Biol.* 18 (12), S. 6897–6909.

Smith, Jason A.; Waldman, Barbara Criscuolo; Waldman, Alan S. (2005): A role for DNA mismatch repair protein Msh2 in error-prone double-strand-break repair in mammalian chromosomes. *Genetics* 170 (1), S. 355–363.

Sommer, L. A. M.; Schaad, M.; Dames, S. A. (2013): NMR- and Circular Dichroism-monitored Lipid Binding Studies Suggest a General Role for the FATC Domain as Membrane Anchor of Phosphatidylinositol 3-Kinase-related Kinases (PIKK). *Journal of Biological Chemistry* 288 (27), S. 20046–20063.

Song, J. (2005): Small Ubiquitin-like Modifier (SUMO) Recognition of a SUMO Binding Motif: A REVERSAL OF THE BOUND ORIENTATION. *Journal of Biological Chemistry* 280 (48), S. 40122–40129.

Sono, A.; Sakaguchi, K. (1988): Isolation and characterization of Chinese hamster cell line resistant to monofunctional alkylating agents. *Somat. Cell Mol. Genet.* 14 (4), S. 329–344.

Spinaci, M.; Vallorani, C.; Bucci, D.; Tamanini, C.; Porcu, E.; Galeati, G. (2012): Vitrification of pig oocytes induces changes in histone H4 acetylation and histone H3 lysine 9 methylation (H3K9). *Vet. Res. Commun.* 36 (3), S. 165–171.

Stanford, W. L.; Cohn, J. B.; Cordes, S. P. (2001): Gene-trap mutagenesis: past, present and beyond. *Nat. Rev. Genet.* 2 (10), S. 756–768.

Stankiewicz, Marta; Nikolay, Rainer; Rybin, Vladimir; Mayer, Matthias P. (2010): CHIP participates in protein triage decisions by preferentially ubiquitinating Hsp70-bound substrates. *FEBS J.* 277 (16), S. 3353–3367.

Stellas, Dimitris; El Hamidieh, Avraam; Patsavoudi, Evangelia (2010): Monoclonal antibody 4C5 prevents activation of MMP2 and MMP9 by disrupting their interaction with extracellular HSP90 and inhibits formation of metastatic breast cancer cell deposits. *BMC Cell Biol.* 11, S. 51.

Stepanov, G. A.; Semenov, D. V.; Kuligina, E. V.; Koval, O. A.; Rabinov, I. V.; Kit, Y. Y.; Richter, V. A. (2012): Analogues of Artificial Human Box C/D Small Nucleolar RNA As Regulators of Alternative Splicing of a pre-mRNA Target. *Acta Naturae* 4 (1), S. 32–41.

Stepanov, Grigoriy A.; Semenov, Dmitry V.; Savelyeva, Anna V.; Kuligina, Elena V.; Koval, Olga A.; Rabinov, Igor V.; Richter, Vladimir A. (2013): Artificial box C/D RNAs affect pre-mRNA maturation in human cells. *Biomed Res Int* 2013, S. 656158.

Sung, Yun Hee; Lee, Jae Seong; Park, Soon Hye; Koo, Jane; Lee, Gyun Min (2007): Influence of co-down-regulation of caspase-3 and caspase-7 by siRNAs on sodium butyrate-induced apoptotic cell death of Chinese hamster ovary cells producing thrombopoietin. *Metab. Eng.* 9 (5-6), S. 452–464.

Sunley, Kevin; Tharmalingam, Tharmala; Butler, Michael (2008): CHO cells adapted to hypothermic growth produce high yields of recombinant beta-interferon. *Biotechnol. Prog.* 24 (4), S. 898–906.

Sun, Zheng; Wu, Tongde; Zhao, Fei; Lau, Alexandria; Birch, Christina M.; Zhang, Donna D. (2011): KPNA6 (Importin {alpha}7)-mediated nuclear import of Keap1 represses the Nrf2-dependent antioxidant response. *Mol. Cell. Biol.* 31 (9), S. 1800–1811.

Tabuchi, Hisahiro; Sugiyama, Tomoya (2013): Cooverexpression of alanine aminotransferase 1 in Chinese hamster ovary cells overexpressing taurine transporter further stimulates metabolism and enhances product yield. *Biotechnol. Bioeng.*

Tabuchi, Hisahiro; Sugiyama, Tomoya; Tanaka, Saeko; Tainaka, Satoshi (2010): Overexpression of taurine transporter in Chinese hamster ovary cells can enhance cell viability and product yield, while promoting glutamine consumption. *Biotechnol. Bioeng.* 107 (6), S. 998–1003.

Taiyab, Aftab; Rao, Ch Mohan (2011): HSP90 modulates actin dynamics: Inhibition of HSP90 leads to decreased cell motility and impairs invasion. *Biochimica et Biophysica Acta (BBA) – Molecular Cell Research* 1813 (1), S. 213–221.

Takagi, M.; Ilias, M.; Yoshida, T. (2000): Selective retension of active cells employing low centrifugal force at the medium change during suspension culture of Chinese hamster ovary cells producing tPA. *J. Biosci. Bioeng.* 89 (4), S. 340–344.

Takano, Keizo; Miki, Takashi; Katahira, Jun; Yoneda, Yoshihiro (2007): NXF2 is involved in cytoplasmic mRNA dynamics through interactions with motor proteins. *Nucleic Acids Res.* 35 (8), S. 2513–2521.

Tan, Ee Phie; Caro, Sarah; Potnis, Anish; Lanza, Christopher; Slawson, Chad (2013): O-linked N-acetylglucosamine cycling regulates mitotic spindle organization. *J. Biol. Chem.* 288 (38), S. 27085–27099.

Tang, Jen-Ruey; Michaelis, Katherine A.; Nozik-Grayck, Eva; Seedorf, Gregory J.; Hartman-Filson, Marlena; Abman, Steven H.; Wright, Clyde J. (2013): The NF-κB inhibitory proteins IκBα and IκBβ mediate disparate responses to inflammation in fetal pulmonary endothelial cells. *J. Immunol.* 190 (6), S. 2913–2923.

Tan, Guangyun; Niu, Jixiao; Shi, Yuling; Ouyang, Hongsheng; Wu, Zhao-Hui (2012): NF-κB-dependent microRNA-125b up-regulation promotes cell survival by targeting p38α upon ultraviolet radiation. *J. Biol. Chem.* 287 (39), S. 33036–33047.

Tan, Hong Kiat; Lee, May May; Yap, Miranda G. S.; Wang, Daniel I. C. (2008): Overexpression of cold-inducible RNA-binding protein increases interferon-gamma production in Chinese-hamster ovary cells. *Biotechnol. Appl. Biochem.* 49 (Pt 4), S. 247–257.

Tani, Hidenori; Torimura, Masaki; Akimitsu, Nobuyoshi (2013): The RNA degradation pathway regulates the function of GAS5 a non-coding RNA in mammalian cells. *PLoS ONE* 8 (1), S. e55684.

Tao, Jun; Lu, Qiang; Wu, Deyao; Li, Pengchao; Xu, Bin; Qing, Weijie *et al.* (2011): microRNA-21 modulates cell proliferation and sensitivity to doxorubicin in bladder cancer cells. *Oncol. Rep.* 25 (6), S. 1721–1729.

Taschwer, Michael; Hackl, Matthias; Hernández Bort, Juan A.; Leitner, Christian; Kumar, Niraj; Puc, Urszula *et al.* (2012): Growth, productivity and protein glycosylation in a CHO EpoFc producer cell line adapted to glutamine-free growth. *J. Biotechnol.* 157 (2), S. 295–303.

Taylor, William Ramsay (1986): The classification of amino acid conservation. *Journal of Theoretical Biology* 119 (2), S. 205–218.

Tey, B. T.; Singh, R. P.; Piredda, L.; Piacentini, M.; Al-Rubeai, M. (2000): Influence of bcl-2 on cell death during the cultivation of a Chinese hamster ovary cell line expressing a chimeric antibody. *Biotechnol. Bioeng.* 68 (1), S. 31–43.

Thompson, L. H. (1979): Mutant isolation. *Meth. Enzymol.* 58, S. 308–322.

Tigges, Marcel; Fussenegger, Martin (2006): Xbp1-based engineering of secretory capacity enhances the productivity of Chinese hamster ovary cells. *Metabolic Engineering* 8 (3), S. 264–272.

Tirmenstein, M. A.; Nicholls-Grzemski, F. A.; Zhang, J. G.; Fariss, M. W. (2000): Glutathione depletion and the production of reactive oxygen species in isolated hepatocyte suspensions. *Chem. Biol. Interact.* 127 (3), S. 201–217.

Traina, Fabíola; Favaro, Patricia M. B.; Medina, Samuel de Souza; Duarte, Adriana da Silva Santos; Winnischofer, Sheila Maria Brochado; Costa, Fernando F.; Saad, Sara T. O. (2006): ANKHD1, ankyrin repeat and KH domain containing 1, is overexpressed in acute leukemias and is associated with SHP2 in K562 cells. *Biochim. Biophys. Acta* 1762 (9), S. 828–834.

Trapnell, Cole; Pachter, Lior; Salzberg, Steven L. (2009): TopHat: discovering splice junctions with RNA-Seq. *Bioinformatics* 25 (9), S. 1105–1111.

Trummer, Evelyn; Ernst, Wolfgang; Hesse, Friedemann; Schriebl, Kornelia; Lattenmayer, Christine; Kunert, Renate *et al.* (2008): Transcriptional profiling of phenotypically different Epo-Fc expressing CHO clones by cross-species microarray analysis. *Biotechnol J* 3 (7), S. 924–937.

Tsukita, S.; Oishi, K.; Akiyama, T.; Yamanashi, Y.; Yamamoto, T. (1991): Specific proto-oncogenic tyrosine kinases of src family are enriched in cell-to-cell adherens junctions where the level of tyrosine phosphorylation is elevated. *J. Cell Biol.* 113 (4), S. 867–879.

Urlaub, G.; Chasin, L. A. (1980): Isolation of Chinese hamster cell mutants deficient in dihydrofolate reductase activity. *Proc. Natl. Acad. Sci. U.S.A.* 77 (7), S. 4216–4220.

Urlaub, G.; Käs, E.; Carothers, A. M.; Chasin, L. A. (1983): Deletion of the diploid dihydrofolate reductase locus from cultured mammalian cells. *Cell* 33 (2), S. 405–412.

Urlaub, G.; Landzberg, M.; Chasin, L. A. (1981): Selective killing of methotrexate-resistant cells carrying amplified dihydrofolate reductase genes. *Cancer Res.* 41 (5), S. 1594–1601.

Ustek, Duran; Sirma, Sema; Gumus, Ergun; Arikan, Muzaffer; Cakiris, Aris; Abaci, Neslihan *et al.* (2012): A genome-wide analysis of lentivector integration sites using targeted sequence capture and next generation sequencing technology. *Infect. Genet. Evol.* 12 (7), S. 1349–1354.

van Durme, Joost; Maurer-Stroh, Sebastian; Gallardo, Rodrigo; Wilkinson, Hannah; Rousseau, Frederic; Schymkowitz, Joost; Tramontano, Anna (2009): Accurate Prediction of DnaK-Peptide Binding via Homology Modelling and Experimental Data. *PLoS Comput Biol* 5 (8), S. e1000475.

van Nues, Robert Willem; Granneman, Sander; Kudla, Grzegorz; Sloan, Katherine Elizabeth; Chicken, Matthew; Tollervey, David; Watkins, Nicholas James (2011): Box C/D snoRNP catalysed methylation is aided by additional pre-rRNA base-pairing. *EMBO J* 30 (12), S. 2420–2430.

Vaute, Olivier; Nicolas, Estelle; Vandel, Laurence; Trouche, Didier (2002): Functional and physical interaction between the histone methyl transferase Suv39H1 and histone deacetylases. *Nucleic Acids Res.* 30 (2), S. 475–481.

Vespa, Alisa; D'Souza, Sudhir J. A.; Dagnino, Lina (2005): A novel role for integrin-linked kinase in epithelial sheet morphogenesis. *Mol. Biol. Cell* 16 (9), S. 4084–4095.

Villeneuve, Louisa M.; Kato, Mitsuo; Reddy, Marpadga A.; Wang, Mei; Lanting, Linda; Natarajan, Rama (2010): Enhanced levels of microRNA-125b in vascular smooth muscle cells of diabetic db/db mice lead to increased inflammatory gene expression by targeting the histone methyltransferase Suv39h1. *Diabetes* 59 (11), S. 2904–2915.

Vitali, Patrice; Basyuk, Eugenia; Le Meur, Elodie; Bertrand, Edouard; Muscatelli, Françoise; Cavaillé, Jérôme; Huttenhofer, Alexander (2005): ADAR2-mediated editing of RNA substrates in the nucleolus is inhibited by C/D small nucleolar RNAs. *J. Cell Biol.* 169 (5), S. 745–753.

Vrba, Lukas; Muñoz-Rodríguez, José L.; Stampfer, Martha R.; Futscher, Bernard W. (2013): miRNA gene promoters are frequent targets of aberrant DNA methylation in human breast cancer. *PLoS ONE* 8 (1), S. e54398.

Wagner, Elizabeth D.; Anderson, Diana; Dhawan, Alok; Rayburn, A. Lane; Plewa, Michael J. (2003): Evaluation of EMS-induced DNA damage in the single cell gel electrophoresis (Comet) assay and with flow cytometric analysis of micronuclei. *Teratog., Carcinog. Mutagen.* 2, S. 1–11.

Waldren, C.; Jones, C.; PUCK, T. T. (1979): Measurement of mutagenesis in mammalian cells. *Proc. Natl. Acad. Sci. U.S.A.* 76 (3), S. 1358–1362.

Walser, J.-C; Furano, A. V. (2010): The mutational spectrum of non-CpG DNA varies with CpG content. *Genome Research* 20 (7), S. 875–882.

Wang, Chiung-Min; Liu, Runhua; Wang, Lizhong; Yang, Wei-Hsiung (2013a): Acidic residue Glu199 increases SUMOylation level of nuclear hormone receptor NR5A1. *Int J Mol Sci* 14 (11), S. 22331–22345.

Wang, Jing; Hevi, Sarah; Kurash, Julia K.; Lei, Hong; Gay, Frédérique; Bajko, Jeffrey *et al.* (2009a): The lysine demethylase LSD1 (KDM1) is required for maintenance of global DNA methylation. *Nat. Genet.* 41 (1), S. 125–129.

Wang, K.; Knipfer, M.; Huang, Q. Q.; van Heerden, A.; Hsu, L. C.; Gutierrez, G. *et al.* (1996): Human skeletal muscle nebulin sequence encodes a blueprint for thin filament architecture. Sequence motifs and affinity profiles of tandem repeats and terminal SH3. *J. Biol. Chem.* 271 (8), S. 4304–4314.

Wang, Lanlan; Wang, Changjun; Su, Bingnan; Song, Quansheng; Zhang, Yingmei; Luo, Yang *et al.* (2013b): Recombinant human PDCD5 protein enhances chemosensitivity of breast cancer in vitro and in vivo. *Biochem. Cell Biol.* 91 (6), S. 526–531.

Wang, Li; Li, Ling; Zhang, Hailong; Luo, Xiao; Dai, Jingquan; Zhou, Shaolian *et al.* (2011): Structure of human SMYD2 protein reveals the basis of p53 tumor suppressor methylation. *J. Biol. Chem.* 286 (44), S. 38725–38737.

Wang, Na; Zhang, Chao-Qi; He, Jia-Huan; Duan, Xiao-Fei; Wang, Yuan-Yuan; Ji, Xiang *et al.* (2013c): miR-21 Down-Regulation Suppresses Cell Growth, Invasion and Induces Cell Apoptosis by Targeting FASL, TIMP3, and RECK Genes in Esophageal Carcinoma. *Dig. Dis. Sci.*

Wang, Peitao; Shu, Zhiquan; He, Liqun; Cui, Xiangdong; Wang, Yuzhen; Gao, Dayong (2005): The pertinence of expression of heat shock proteins (HSPs) to the efficacy of cryopreservation in HELAs. *Cryo Letters* 26 (1), S. 7–16.

Wang, Qing; Song, Changcheng; Li, Chou-Chi H. (2004): Molecular perspectives on p97-VCP: progress in understanding its structure and diverse biological functions. *J. Struct. Biol.* 146 (1-2), S. 44–57.

Wang, X.; Zhu, K.; Li, S.; Liao, Y.; Du, R.; Zhang, X. *et al.* (2012): MLL1, a H3K4 methyltransferase, regulates the TNF -stimulated activation of genes downstream of NF- B. *Journal of Cell Science* 125 (17), S. 4058–4066.

Wang, Zhong; Gerstein, Mark; Snyder, Michael (2009b): RNA-Seq: a revolutionary tool for transcriptomics. *Nat. Rev. Genet.* 10 (1), S. 57–63.

Webster, G. A.; Perkins, N. D. (1999): Transcriptional cross talk between NF-kappaB and p53. *Mol. Cell. Biol.* 19 (5), S. 3485–3495.

Weiss, Frank Ulrich; Marques, Ines J.; Woltering, Joost M.; Vlecken, Danielle H.; Aghdassi, Ali; Partecke, Lars Ivo *et al.* (2009): Retinoic acid receptor antagonists inhibit miR-10a expression and block metastatic behavior of pancreatic cancer. *Gastroenterology* 137 (6), S. 2136-45.e1-7.

Wei, Yi-Yun C.; Naderi, Saeideh; Meshram, Mukesh; Budman, Hector; Scharer, Jeno M.; Ingalls, Brian P.; McConkey, Brendan J. (2011): Proteomics analysis of chinese hamster ovary cells undergoing apoptosis during prolonged cultivation. *Cytotechnology* 63 (6), S. 663–677.

Werner, Andreas; Flotho, Annette; Melchior, Frauke (2012): The RanBP2/RanGAP1*SUMO1/Ubc9 complex is a multisubunit SUMO E3 ligase. *Mol. Cell* 46 (3), S. 287–298.

White, Joshua; Li, Zhihua; Sardana, Richa; Bujnicki, Janusz M.; Marcotte, Edward M.; Johnson, Arlen W. (2008): Bud23 methylates G1575 of 18S rRNA and is required for efficient nuclear export of pre-40S subunits. *Mol. Cell. Biol.* 28 (10), S. 3151–3161.

Whitelaw, C. B.; Archibald, A. L.; Harris, S.; McClenaghan, M.; Simons, J. P.; Clark, A. J. (1991): Targeting expression to the mammary gland: intronic sequences can enhance the efficiency of gene expression in transgenic mice. *Transgenic Res.* 1 (1), S. 3–13.

Williams, Gwyn T.; Mourtada-Maarabouni, Mirna; Farzaneh, Farzin (2011): A critical role for non-coding RNA GAS5 in growth arrest and rapamycin inhibition in human T-lymphocytes. *Biochem. Soc. Trans.* 39 (2), S. 482–486.

Witenberg, B.; Kalir, H. H.; Raviv, Z.; Kletter, Y.; Kravtsov, V.; Fabian, I. (1999): Inhibition by ascorbic acid of apoptosis induced by oxidative stress in HL-60 myeloid leukemia cells. *Biochem. Pharmacol.* 57 (7), S. 823–832.

Wlaschin, Katie F.; Hu, Wei-Shou (2007): Engineering cell metabolism for high-density cell culture via manipulation of sugar transport. *J. Biotechnol.* 131 (2), S. 168–176.

Wong, Daniel; Teixeira, Ana; Oikonomopoulos, Spyros; Humburg, Peter; Lone, Imtiaz; Saliba, David *et al.* (2011): Extensive characterization of NF-κB binding uncovers non-canonical motifs and advances the interpretation of genetic functional traits. *Genome Biol* 12 (7), S. R70.

Wong, E-T; Ngoi, S-M; Lee, C. G. L. (2002): Improved co expression of multiple genes in vectors containing internal ribosome entry sites (IRESes) from human genes. *Gene Ther.* 9 (5), S. 337–344.

Wu, Cheng-Tao; Chiou, Chien-Ying; Chiu, Ho-Chen; Yang, Ueng-Cheng (2013): Fine-tuning of microRNA-mediated repression of mRNA by splicing-regulated and highly repressive microRNA recognition element. *BMC Genomics* 14, S. 438.

Wu, Li-Peng; Wang, Xi; Li, Lian; Zhao, Ying; Lu, Shaoli; Yu, Yu et al. (2008): Histone deacetylase inhibitor depsipeptide activates silenced genes through decreasing both CpG and H3K9 methylation on the promoter. *Mol. Cell. Biol.* 28 (10), S. 3219–3235.

Wu, Shwu-Yuan; Chiang, Cheng-Ming (2009): p53 sumoylation: mechanistic insights from reconstitution studies. *Epigenetics* 4 (7), S. 445–451.

Xiaoguang Liu, Jian Liu Tasha Williams Wright Jessica Lee Peggy Lio Laurel Donahue-Hjelle Paula Ravnikar Florence Wu (2010): Isolation of Novel High-Osmolarity Resistant CHO DG44 Cells After Suspension of DNA Mismatch Repair. *BioProcess Int.* 8 (4), S. 68–76.

Xing, Xinrong; Bi, Hailian; Chang, Alan K.; Zang, Ming-Xi; Wang, Miao; Ao, Xiang et al. (2012): SUMOylation of AhR modulates its activity and stability through inhibiting its ubiquitination. *J. Cell. Physiol.* 227 (12), S. 3812–3819.

Xing, Zizhuo; Li, Zhengjian; Chow, Vincent; Lee, Steven S. (2008): Identifying inhibitory threshold values of repressing metabolites in CHO cell culture using multivariate analysis methods. *Biotechnol. Prog.* 24 (3), S. 675–683.

Xiong, Shudao; Zheng, Yijie; Jiang, Pei; Liu, Ronghua; Liu, Xiaoming; Chu, Yiwei (2011): MicroRNA-7 inhibits the growth of human non-small cell lung cancer A549 cells through targeting BCL-2. *Int. J. Biol. Sci.* 7 (6), S. 805–814.

Xu, Fenglei; Wu, Kai; Zhao, Miaoqing; Qin, Yejun; Xia, Ming (2013): Expression and clinical significance of the programmed cell death 5 gene and protein in laryngeal squamous cell carcinoma. *J. Int. Med. Res.* 41 (6), S. 1838–1847.

Xu, Lanjun; Hu, Jing; Zhao, Yuanbo; Hu, Jia; Xiao, Juan; Wang, Yanming et al. (2012a): PDCD5 interacts with p53 and functions as a positive regulator in the p53 pathway. *Apoptosis* 17 (11), S. 1235–1245.

Xu, Xiao-Yin; Wang, Wen-Qian; Zhang, Lei; Li, Yi-Ming; Tang, Miao; Jiang, Nan et al. (2012b): Clinical implications of p57 KIP2 expression in breast cancer. *Asian Pac. J. Cancer Prev.* 13 (10), S. 5033–5036.

Xu, Xun; Nagarajan, Harish; Lewis, Nathan E.; Pan, Shengkai; Cai, Zhiming; Liu, Xin et al. (2011): The genomic sequence of the Chinese hamster ovary (CHO)-K1 cell line. *Nat. Biotechnol.* 29 (8), S. 735–741.

Yang, Shen-Hsi; Galanis, Alex; Witty, James; Sharrocks, Andrew D. (2006): An extended consensus motif enhances the specificity of substrate modification by SUMO. *EMBO J.* 25 (21), S. 5083–5093.

Yang, Xiao-Dong; Chen, Lin-Feng (2011): Talking to histone: methylated RelA serves as a messenger. *Cell Res.* 21 (4), S. 561–563.

Yang, Yuansheng; Mariati; Chusainow, Janet; Yap, Miranda G.S (2010): DNA methylation contributes to loss in productivity of monoclonal antibody-producing CHO cell lines. *Journal of Biotechnology* 147 (3-4), S. 180–185.

Yee, Joon Chong; Gerdtzen, Ziomara P.; Hu, Wei-Shou (2009): Comparative transcriptome analysis to unveil genes affecting recombinant protein productivity in mammalian cells. *Biotechnol. Bioeng.* 102 (1), S. 246–263.

Yee, Joon Chong; Wlaschin, Katie Fraass; Chuah, Song Hui; Nissom, Peter Morin; Hu, Wei-Shou (2008): Quality assessment of cross-species hybridization of CHO transcriptome on a mouse DNA oligo microarray. *Biotechnol. Bioeng.* 101 (6), S. 1359–1365.

Yokobori, Takehiko; Yokoyama, Yozo; Mogi, Akira; Endoh, Hideki; Altan, Bolag; Kosaka, Takayuki et al. (2014): FBXW7 Mediates Chemotherapeutic Sensitivity and Prognosis in NSCLCs. *Mol. Cancer Res.* 12 (1), S. 32–37.

Yonemura, S.; Hirao, M.; Doi, Y.; Takahashi, N.; Kondo, T.; Tsukita, S. (1998): Ezrin/radixin/moesin (ERM) proteins bind to a positively charged amino acid cluster in the juxta-membrane cytoplasmic domain of CD44, CD43, and ICAM-2. *J. Cell Biol.* 140 (4), S. 885–895.

Yoon, Sung Kwan; Ahn, Yong-Ho; Jeong, Myeong Hyeon (2007): Effect of culture temperature on follicle-stimulating hormone production by Chinese hamster ovary cells in a perfusion bioreactor. *Appl. Microbiol. Biotechnol.* 76 (1), S. 83–89.

Yoon, Sung Kwan; Hong, Jong Kwang; Choo, Seung Ho; Song, Ji Yong; Park, Hong Woo; Lee, Gyun Min (2006): Adaptation of Chinese hamster ovary cells to low culture temperature: cell growth and recombinant protein production. *J. Biotechnol.* 122 (4), S. 463–472.

Young, Jason C.; Hoogenraad, Nicholas J.; Hartl, F. Ulrich (2003): Molecular chaperones Hsp90 and Hsp70 deliver preproteins to the mitochondrial import receptor Tom70. *Cell* 112 (1), S. 41–50.

Yu, Chao; Crispin, Max; Sonnen, Andreas F-P; Harvey, David J.; Chang, Veronica T.; Evans, Edward J. *et al.* (2011): Use of the α-mannosidase I inhibitor kifunensine allows the crystallization of apo CTLA-4 homodimer produced in long-term cultures of Chinese hamster ovary cells. *Acta Crystallogr. Sect. F Struct. Biol. Cryst. Commun.* 67 (Pt 7), S. 785–789.

Yu, Jingwei; Zhang, Yingsha; Qi, Zhongxia; Kurtycz, Daniel; Vacano, Guido; Patterson, David (2008): Methylation-mediated downregulation of the B-cell translocation gene 3 (BTG3) in breast cancer cells. *Gene Expr.* 14 (3), S. 173–182.

Yun, Chee Yong; Liu, Sen; Lim, Sing Fee; Wang, Tianhua; Chung, Beatrice Y. F.; Jiat Teo, Joong *et al.* (2007): Specific inhibition of caspase-8 and -9 in CHO cells enhances cell viability in batch and fed-batch cultures. *Metab. Eng.* 9 (5-6), S. 406–418.

Yu, Y. T.; Shu, M. D.; Steitz, J. A. (1998): Modifications of U2 snRNA are required for snRNP assembly and pre-mRNA splicing. *EMBO J.* 17 (19), S. 5783–5795.

Zachara, Natasha E.; Molina, Henrik; Wong, Ker Yi; Pandey, Akhilesh; Hart, Gerald W. (2011): The dynamic stress-induced "O-GlcNAc-ome" highlights functions for O-GlcNAc in regulating DNA damage/repair and other cellular pathways. *Amino Acids* 40 (3), S. 793–808.

Zanghi, J. A.; Mendoza, T. P.; Knop, R. H.; Miller, W. M. (1998): Ammonia inhibits neural cell adhesion molecule polysialylation in Chinese hamster ovary and small cell lung cancer cells. *J. Cell. Physiol.* 177 (2), S. 248–263.

Zeng, Yu; Kulkarni, Prakash; Inoue, Takahiro; Getzenberg, Robert H. (2009): Down-regulating cold shock protein genes impairs cancer cell survival and enhances chemosensitivity. *J. Cell. Biochem.* 107 (1), S. 179–188.

Zeng, Yu; Wodzenski, Dana; Gao, Dong; Shiraishi, Takumi; Terada, Naoki; Li, Youqiang *et al.* (2013): Stress-response protein RBM3 attenuates the stem-like properties of prostate cancer cells by interfering with CD44 variant splicing. *Cancer Res.* 73 (13), S. 4123–4133.

Zeyda, M.; Borth, N.; Kunert, R.; Katinger, H. (1999): Optimization of sorting conditions for the selection of stable, high-producing mammalian cell lines. *Biotechnol. Prog.* 15 (5), S. 953–957.

Zhang, N.; Li, X.; Wu, C. W.; Dong, Y.; Cai, M.; Mok, M. T. S. *et al.* (2012): microRNA-7 is a novel inhibitor of YY1 contributing to colorectal tumorigenesis. *Oncogene*.

Zhang, Z.; Zhu, Z.; Watabe, K.; Zhang, X.; Bai, C.; Xu, M. *et al.* (2013): Negative regulation of lncRNA GAS5 by miR-21. *Cell Death Differ.* 20 (11), S. 1558–1568.

Zhao, Ruiying; Yang, Heng-Yin; Shin, Jihyun; Phan, Liem; Fang, Lekun; Che, Ting-Fang *et al.* (2013a): CDK inhibitor p57 (Kip2) is downregulated by Akt during HER2-mediated tumorigenicity. *Cell Cycle* 12 (6), S. 935–943.

Zhao, Xinliang; Li, Zhu-Hong; Terns, Rebecca M.; Terns, Michael P.; Yu, Yi-Tao (2002): An H/ACA guide RNA directs U2 pseudouridylation at two different sites in the branchpoint recognition region in Xenopus oocytes. *RNA* 8 (12), S. 1515–1525.

Zhao, Xue-Ming; Ren, Jing-Jing; Du, Wei-Hua; Hao, Hai-Sheng; Wang, Dong; Qin, Tong *et al.* (2013b): Effect of vitrification on promoter CpG island methylation patterns and expression levels of DNA methyltransferase 1o, histone acetyltransferase 1, and deacetylase 1 in metaphase II mouse oocytes. *Fertility and Sterility* 100 (1), S. 256–261.

Zhou, Meixia; Crawford, Yongping; Ng, Domingos; Tung, Jack; Pynn, Abigail F. J.; Meier, Angela *et al.* (2011): Decreasing lactate level and increasing antibody production in Chinese Hamster Ovary cells (CHO) by reducing the expression of lactate dehydrogenase and pyruvate dehydrogenase kinases. *J. Biotechnol.* 153 (1-2), S. 27–34.

Zhou, Rui; Hu, Guoku; Gong, Ai-Yu; Chen, Xian-Ming (2010): Binding of NF-kappaB p65 subunit to the promoter elements is involved in LPS-induced transactivation of miRNA genes in human biliary epithelial cells. *Nucleic Acids Res.* 38 (10), S. 3222–3232.

Zhu, D.; Bourguignon, L. Y. (1998): The ankyrin-binding domain of CD44s is involved in regulating hyaluronic acid-mediated functions and prostate tumor cell transformation. *Cell Motil. Cytoskeleton* 39 (3), S. 209–222.

Zhu, Xiaolan; Li, Yuefeng; Shen, Huiling; Li, Hao; Long, Lulu; Hui, Lulu; Xu, Wenlin (2013): miR-137 inhibits the proliferation of lung cancer cells by targeting Cdc42 and Cdk6. *FEBS Lett.* 587 (1), S. 73–81.

Zhu, Z-H; Yu, Y. P.; Shi, Y-K; Nelson, J. B.; Luo, J-H (2009): CSR1 induces cell death through inactivation of CPSF3. *Oncogene* 28 (1), S. 41–51.

Zientek-Targosz, Helena; Kunnev, Dimiter; Hawthorn, Lesleyann; Venkov, Mikhail; Matsui, Sei-Ichi; Cheney, Richard T.; Ionov, Yuri (2008): Transformation of MCF-10A cells by random mutagenesis with frameshift mutagen ICR191: a model for identifying candidate breast-tumor suppressors. *Mol. Cancer* 7, S. 51.

Zuker, Michael (2003): Mfold web server for nucleic acid folding and hybridization prediction. *Nucleic Acids Res.* 31 (13), S. 3406–3415.

8 Abbreviations

AscH	Ascorbic acid	lncRNA	Long non-conding RNA
BFT50	Bioreactor filter tube 50	mAb	monoclonal antibody
cgr	*Cricetulus griseus*	miRNA	microRNA
CHO	Chinese hamster ovary	*mmu*	*Mus musculus*
Csp	Cell specific productivity	$n_{biological}$	biological replicate number
EF	CD CHO EfficientFeed A/B	NGS	Next-generation sequencing
EMS	Ethylmethane sulfonate	*qRT-PCR*	quantitative real-time polymerase chain reaction
FC	Fold change	*rno*	*Rattus norvegicus*
Gln	L-Glutamine	Snora	H/ACA box snoRNA
HCD	High cell density	Snord	C/D box snoRNA
hIgG	humanised *immu*noglobulin G	snoRNA	small nucleolar RNA
HyOsm	Selected at high osmolalities	snRNA	small nuclear RNA
$IRES_{FMDV}$	Internal ribosomal entry site derived from foot-and-mouth disease virus	tcd	total cell density
IS	IS CHO Feed-CD XP	TPR	Tetratricopeptide repeat
Ivcd	integral viable cell density	wgs	whole genome shotgun sequence
LD	Limited dilution	vcd	viable cell density

Bisher erschienene Bände der Reihe

Bielefelder Schriften zur molekularen Biotechnologie

ISSN: 2364-4877

Vormals "Bielefelder Schriften zur Zellkulturtechnik"
(ISSN: 1866-9727, Vol. 1-13)

9	Benjamin Müller	Differentielle Analyse des Phosphoproteoms Apoptose-induzierter Jurkat ACC 282-Zellen
		ISBN 978-3-8325-3476-9 38.00 €
10	Sandra Klausing	Optimierung von CHO Produktionszelllinien: RNAi-vermittelter Gen-*knockdown* und Untersuchungen zur Klonstabilität
		ISBN 978-3-8325-3594-0 55.40 €
11	Eva Skerhutt	Proteomanalysen zur Aufklärung regulatorischer Prozesse im Zentralstoffwechsel der humanen Produktionszelllinie AGE1.HN
		ISBN 978-3-8325-3653-4 53.30 €
12	Jennifer Becker-Strugholtz	Transkriptomsequenzierung von CHO-Zelllinien zur Entwicklung eines spezifischen Microarrays für die Analyse des Einflusses wachstumsfördernder Substanzen
		ISBN 978-3-8325-3657-2 49.50 €
13	Christina Timmermann	Transkriptomanalyse von rekombinanten CHO-Zellen. Von der Microarray-Entwicklung bis zur Charakterisierung von Fermentationsprozessen
		ISBN 978-3-8325-3770-8 36.50 €
...	....	....
14	R. S. Velur Selvamani	Continuous culture and extracellular recombinant protein expression in *Escherichia coli*
		ISBN 978-3-8325-3951-1 50.00 €
15	Beat Thalmann	Identification of novel modulators towards high cell density and high-producing Chinese hamster ovary suspension cell cultures as well as their application in biopharmaceutical protein production
		ISBN 978-3-8325-4046-3 68.50 €

Alle erschienenen Bücher können unter der angegebenen ISBN im Buchhandel oder direkt beim Logos Verlag Berlin (www.logos-verlag.de, Fax: 030 42 85 10 92) bestellt werden.